THE UNTRAINED ENVIRONMENTALIST

How an Australian grazier brought his barren property back to life

JOHN FENTON

ALLEN&UNWIN

First published in 2010

Allen & Unwin
83 Alexander Street
Crows Nest NSW 2065
Australia
Phone: (61 2) 8425 0100
Fax: (61 2) 9906 2218
Email: info@allenandunwin.com
Web: www.allenandunwin.com

Cataloguing-in-Publication details are available
from the National Library of Australia
www.librariesaustralia.nla.gov.au

ISBN 978 1 74237 019 4

Internal design by Phil Campbell
Map by Ian Faulkner
Set in 12/16 pt Janson Text LT Std by Bookhouse, Sydney

10 9 8 7 6 5 4 3 2 1

For my wife of 50 years, Cicely, and our very close family: Amanda (and Charles), David (and Cath), Patrick (and Bronnie) and John, plus our grandchildren, who have brought so much joy to our lives—Clive, Anna, Lucy, Will, Harriet, Jock, Georgia and Tanner

CONTENTS

FOREWORD

By Graeme Gunn

The father did not leave a forest for the boy,
so the boy sowed his own seed and grew his own forest,
and as the forest grew, the boy grew until he became a man—
a good man.

We were sitting on the terrace at Lanark one evening as the twilight sun hit the eucalypt plantation on the far side of Lake Cicely, when successive flocks of ibis crossed the sky heading south. We counted 6000 birds by our reckoning—it was an armada, an awesome and remarkable sight. One had to assume that the waters of Lanark were a beacon for that night crossing.

John Fenton and I had grown up in Hamilton, the centre of a productive wool-growing area in the Western District of Victoria. We had known each other but did not establish a friendship until 1954 when we were in our early twenties; both keen to leave our mark on the world but without any apparent skills. At that time I was impressed by his self-effacing eccentricities and his principled and romantic approach to finding a worthwhile position for his own life. In answer to my questions about where he had been and what he was currently working on, he described his time away at boarding school in Geelong, his next educational adventure at Longerenong Agricultural College in central Victoria, and his removal from there with others for a minor drinking misdemeanour. He then told me of his most recent job, as a jackaroo on a significant grazing property in central Victoria. When asked what his tasks were as a jackaroo, he replied, 'Picking up sticks.' I thought this both very funny and engagingly self-deprecating. We became solid friends and over the following 50 years have assisted each other and our extended families, intervened in each other's lives and generally had a lot to talk about while extending each other's knowledge.

I well remember in those early days in Hamilton how high and sometimes naïve our aspirations were, and in retrospect, what little personal substance we had to realise them. During this early period while we were both still in Hamilton, I had bought a burnt-out MG car to hopefully restore to its proper state of glory. We spent nights grinding, scraping and sanding metal frames and panels, interspersed with the application of self-generated philosophical graffiti to the walls of the shed with fingers dipped in gelatinous grease.

The resurrection of the car succumbed to other pressures. I traded the carburetors I had bought to fund my move to Melbourne to study architecture and John focused on the embryonic farming property he had inherited near Branxholme, a small hamlet some 25 kilometres from Hamilton.

John had inherited the property from his father, who died when John was young, and his move into farming was consequently different from that of most of his peers. They had grown up on established grazing properties working under parental regimes and conservative farming practices that gave little credence to land care, conservation and the benefits of integrated environments.

John started with a windswept grassland property generally devoid of trees and endowed with some rudimentary structures: two small cottages, a small shearing shed and a small antiquated grain and tool shed. A row of semi-mature pines lined the northern road frontage of the property but these, together with a small stand of sugar gums near the sheds, were not a great defence against the prevailing cold southwesterlies that ripped across the paddocks to chill the spine and rock the cottage.

There were some advantages in taking over such an undeveloped rural canvas. John faced a lot of challenges but these could be met without the trappings of established conservative practices and hierarchal authority. The dreams of creating a farming enterprise that balanced good farming practice with conservation and environmental programs to enhance the rural landscape were part of a distant horizon, but they were realisable, personal dreams, possible of being achieved without interference. This rather lonely pursuit, while not easy, developed in John strong independent characteristics that were assisted and balanced by his lovely, suitably rational soul mate, Cicely.

I have spent many hours at Lanark over the years on my return trips to Hamilton and saw each of the innovative transformations which were applied to this once bare portion of rural land. It was always a delight to see a new water body or a new plantation, initially as habitat and shelter belt and later as part of the farm's productivity program extending the landscape and realising the dream.

During the 1950s and early 1960s, sheep farming in the Western District reached a peak, aided by the Korean War and developing international markets. Then, following a major drought in 1967 and the advancing competitive quality of synthetic fibres, the wool industry suffered a dramatic decline. The damage caused by the drought and the slowness of landholders to adopt more appropriate techniques to land management had been significant. The Fentons' early reintroduction of water bodies surrounding their farmhouse helped to generate a new approach to droughtproofing, fire protection, stock sustenance; all integrated with an evolving natural landscape.

Their farm developed in response to a new sensitivity and a perception of land use that will be necessary for the future wellbeing of the land, of farming practices and also for those involved in the process.

That the Lanark experience has occurred without major support from government authorities is remarkable and signifies the character and personal resources of the whole Fenton family. It has not only been the lack of external funding that marks the Lanark experience and the Fenton family. For many years their different and innovative land-use practices were ridiculed by a sceptical local farming community, and the adage of a prophet not being accepted in his own country could well apply to John Fenton and his family.

The benefits that have accrued due to the Lanark experiment have proved more far-reaching than just the Fentons and the environment they helped create. An extended group of experts and interested bodies have added to and taken from the Fentons and their program. The acceptance of visiting students and academics who have shared, regularly, the unique teaching qualities of John Fenton have left Lanark more aware and better prepared to deliver their own environmentally infused projects. John's work in this field culminated in his receipt of an honorary doctorate in landscape architecture from RMIT University.

The Fentons belong to a small group of innovators who through their perception, integrity and persistence, have proved there is a better, more appropriate way of balancing modern farming practices with conservation and environmental objectives to effect a more sustainable rural Australian landscape for the benefit of all.

Graeme Gunn, LF AIA,
Honorary Doctorate in Architecture RMIT University,
Adjunct Professor in Architecture RMIT University
1 February 2010

ACKNOWLEDGMENTS

Numerous people have contributed to this book in different ways and at different times, and I am deeply indebted to all of them. In particular, I would like to thank Rochelle Ruddock, on whose outstanding research much of the book is based; Murray Gunn, for his 60 years of productive involvement at Lanark; Elizabeth Jacka, author of the most authoritative study of Lanark's birds yet conducted; Terri and Dale Lloyd, whose success at restoring the environment on their own farm has been an inspiration to us all; Bill Middleton OAM, forester and a pioneering advocate of revegetating rural Australia, whose excellent memory I tapped several times while writing this book; Neville Bonney, noted author and botanist, whom I quote in the text several times; Mick Cook and Jim Murphy, whose expert knowledge proved so valuable in the early days at Lanark; Emeritus Professor Jim Sinatra for his continuing support and advice; Roger Young, friend and colleague; Pip Courtney, the esteemed television journalist who first brought the Lanark story to national attention; Dr Rod Bird OAM, retired research scientist and naturalist, who allowed me generous access to his personally compiled reference material; Philip Derriman, who provided expert help with the text; Roz Lawson and Terrie Nicholson for their secretarial skills and friendship; D.J.H. Baker for artistic advice and family background; David Hay, family friend, landscape architect, rural planner and aerial photographer; and Bruce Maloney, family historian, who generously provided both information and photographs. Photos credited to Jim Sinatra and Phin Murphy are from their book, *Listen to the People, Listen to the Land* (Melbourne University Press, 1999). Finally, I must thank my old friend, confidant and lifelong supporter, Adjunct Professor Dr Graeme C. Gunn, known to me always as Gunny, for writing the foreword.

AUTHOR'S NOTE

Throughout this book, I generally refer to trees by their common names. I decided to do this after considerable thought because I realise many ordinary readers would find botanical names rather daunting on the page, even if they were simply included in brackets. At the same time I know that many other readers, especially those with a keen interest in trees, much prefer botanical names. In New Zealand, botanical names are used routinely by everyone involved in forestry. Some years ago a New Zealand forester visited Lanark to have a look at my tree-planting program. He saw a group of trees growing beside the road and asked what they were. I replied, 'Spotted gums.' He said, 'Spotted gums? What are you talking about? I want to know what species they are.' For some reason, I could not remember their botanical name, *Eucalyptus maculata*. He obviously thought he was dealing with an ignoramus and probably lost much of his interest in what I was doing at Lanark.

Since then, for the record, spotted gums have been reclassified as *Corymbia maculata*.

In any case, the use of common names can cause confusion, given that the same common name can apply to quite different types of trees in different parts of Australia. Blue gum is an example. In our pantry there is a jar of blue gum honey that we were given by a nephew who lives just over the border in South Australia, barely an hour's drive away. There, 'blue gum' is the common name for *Eucalyptus leucoxylon*. In Victoria, though, 'blue gum' is *Eucalyptus globulus*, an entirely different tree now grown in vast plantations for woodchips. I can recommend *Eucalyptus leucoxylon* honey, by the way. It is one of the best I have tried.

In spite of all these reservations, however, I have opted to use common names in most cases to avoid clogging the narrative, although the botanical name will also be included in parenthesis in the first reference to each tree or other kind of plant. Moreover, for every common name mentioned in the text, a botanical name is listed in an appendix at the back.

Vasey Farm
Cavendish
THE GRAMPIANS
Ararat
Hamilton
Lanark
Branxholme
Warrnambool
Portland
0
50 km
BASS STRAIT
Mildura
NEW SOUTH WALES
0
150
KILOMETRES
Canberra
ACT
VICTORIA
Echuca
Wodonga
Dimboola
Bendigo
Mansfield
THE GRAMPIANS
Ararat
Ballarat
Melbourne
GREAT DIVIDING RANGE
Mallacoota
Cavendish
Hamilton
Branxholme
Geelong
Sale
Warrnambool
Portland
Portsea
SA
AREA ENLARGED
Western District
Far Western District
BASS STRAIT
TASMAN SEA

1

LANARK'S STORY

Conservation is a state of harmony between men and land. By land is meant all the things on, over, or in the earth. Harmony with land is like harmony with a friend; you cannot cherish his right hand and chop off his left. That is to say, you cannot love game and hate predators; you cannot conserve the waters and waste the ranges; you cannot build the forest and mine the farm. The land is one organism.

Aldo Leopold

Someone in the publishing industry described me recently as an 'accidental environmentalist'. I assume he meant that I acquired my interest in the environment somehow by chance. This is not quite true, although I can see what he was getting at. It is certainly a fact that I do not have a 'green' background: I did not start out with a special interest in the welfare of the environment. I started out as a young farmer who was determined to produce as much fine wool and as many fat wethers and beef cattle as possible on the 1715 acres (nearly

700 hectares) of grazing land that I had at my disposal. Before long, though, I came to see there had to be a compromise, some kind of balance, between a farmer's need for greater and greater output and the land's ability to supply it. In other words, I could see that the land was not an inexhaustible source of production. If you wanted the land to keep producing, you had to care for its needs. You had to give something back.

It was hardly a new concept, but it did start me wondering if the way I was managing my land—which at that time was the way nearly all Australian farmers managed their land—was really the best way. In other words, I began to question what previously I had taken for granted. Gradually, it dawned on me that my 1715 acres of land was really a part of the natural environment—a key realisation, incidentally. Indeed, the closer I looked at the relationship between my land and the environment, the more I realised that you could not separate one from the other. The health of one depended on the health of the other. This was how, bit by bit, I began to develop an interest in the environment.

Lanark, a family property, was a bare, worn-out, windswept place when I took it over as a 21-year-old in 1956. It was obvious that Lanark's land had been badly neglected for many years. The question was: how best to restore the land to good health? The answer that I more or less stumbled on was to start by planting trees. When I say 'stumbled on' I mean that planting trees was not a carefully thought-out solution to my property's problems. It just seemed the right thing to do at the time. To some extent I may have simply been reacting to the fact that the property looked so bare and desolate.

By planting trees I was reversing a process that had been under way across rural Australia ever since Europeans began farming here: clearing the land of trees. A few years ago, according to the Australian Conservation Foundation, more than half a million hectares of land, or well over a million acres, were still being cleared in Australia each year. Australia then had the fifth highest rate of land clearing in the world, higher even than impoverished nations like Burma, Nigeria and the Congo. Today, it seems the rate of land clearing in Australia has declined, thanks largely to an almost total ban on it in Queensland, previously by far the worst offender, which came into effect in 2007.

People who drive through districts like the Wimmera in Victoria must wonder why previous generations of farmers cleared the land so thoroughly. Why did they need to cut down just about every tree that was standing? Why not leave a patch of scrub here and there just for appearance's sake? Or as a windbreak? I am not sure, but I would guess that the farmers who did the clearing simply viewed all trees as obstacles to progress. They would have believed unreservedly in the idea that the more trees they cleared the more productive their farms would be. Let us not forget, too, that cleared land was extremely fertile. Farmers who planted crops in newly cleared land reaped the benefit of thousands of years' accumulation of nutrients in the soil. They did not realise that years of cropping would rapidly exhaust those nutrients.

So my wife Cicely and I started planting trees at Lanark in large numbers. And we kept on planting them, year after year, decade after decade. We also reinstated wetlands to provide a habitat for water birds. Here, too, we were going against the trend. In those days Australian farmers were not only hell-bent on getting rid of trees, they were also madly draining wetlands. They wanted to maximise the area they had available for grazing and cropping, and one simple way to do that—or so they believed—was to drain any land that was too wet for grazing. The ideal farming landscape, as viewed by many Australian farmers, was a bare expanse where grass and crops could grow unhindered. Here and there they may have left clumps of trees to provide shade for the stock, but that was about all. Swamps were got rid of as a matter of course.

Having begun planting trees and native shrubs, we then expanded into agroforestry, by which I mean that many of the trees we planted were grown to be harvested eventually as timber and, accordingly, were pruned regularly. We also set aside 65 acres (26 hectares) of land as a wildlife reserve and planted it out with native trees and shrubs. All this was unconventional farming practice to say the least. In fact, most other farmers regarded me as an eccentric. Yet, despite this and various other difficulties, Cicely and I pressed on, encouraged by the fact that what we were doing was showing results. Environmentally, the property was transformed. Bird life and insect life multiplied at an amazing rate. Our land became healthier, more fertile. Our sheep, too, were healthier. By the time my son David took over the property in 2003, around 20 per cent of the property's 1715 acres were covered by trees—not in large plantations, of course, but mainly in woodlots and rows.

Today, I am satisfied that Lanark is as healthy in an environmental sense as any working farm can be. I am also satisfied that this healthier environment has translated into a more productive farm. The evidence for this is overwhelming. I thought an RMIT (Royal Melbourne Institute of Technology) University publication summed it up well a few years ago. It said: 'Today Lanark provides an excellent demonstration of the value of trees and water on farmland and the balance that can be achieved between agriculture and conservation. The Fentons have created a significant sustainable ecosystem that integrates agriculture, forestry and wildlife.'

Do I think it would be a good idea, therefore, if every Australian farm became as environment-friendly as Lanark? Yes, I do, for the farm's sake as well as for the environment, but this is not to say I would advise every Australian farmer to do what I have done. Having done it, I know better than anyone how hard it has been—physically, financially, even emotionally. If I had been a wealthy businessman who farmed for pleasure, it would have been different, for I could have devoted as much time and money as I liked to environmental work without worrying whether I could afford it. Genuine farmers, though, are already stretched to the limit financially. I believe very few of them would be capable of doing what really needs to be done to protect the environment without going broke. I speak from personal experience here; on one occasion I nearly went broke myself.

So what is to be done? In my view, there is only one option: if Australians want to see their unique and fragile environment cared for and protected, as I am sure most of them do, then the Australian Government has to pay farmers to do it. There is simply no alternative. Farmers are the only people capable of doing the job, and as things stand now most farmers cannot possibly afford to do what needs to be done. A government initiative is essential.

On their side, farmers have to accept that, although their title deeds might show that they own their land, the reality is they are merely custodians of it during their lifetimes. The land was there for millions of years before they were heard of, and it will be there for millions of years after they are forgotten. The American environmentalist Aldo Leopold expressed this thought in his famous 1949 book *A Sand County Almanac*, which I bought in the mid-1960s and have quoted from ever since. He wrote:

> That the land is a community is a basic concept of ecology, but that land is to be loved and respected is an extension of ethics. That land yields a cultural harvest is a fact long known but latterly often forgotten . . . We abuse the land because we regard it as a commodity belonging to us. When we see the land as a community to which we belong, we may begin to see it with love and respect.

Farmers are often reminded that they have a responsibility to look after the land. This is true, although I personally prefer to see it as a privilege rather than a responsibility. This is how Lars-Eric Astrom, a leading figure in Sweden's forestry industry, once described it. He was speaking of forests, but he could just as easily have been speaking of any farm. He said, 'Anyone owning woodland in any country is responsible for a small part of our world. This is a privilege which requires knowledge.'

For the past 200-odd years we Australians have viewed the land as a commodity to be exploited. This attitude, I believe, has been primarily responsible for the degradation of the Australian landscape that is evident everywhere. I am convinced that we must fundamentally change the way we view the land if we are to change the way we manage it. My own belief is that we should view the land as the Aborigines did. They regarded it with reverence, so they treated it with reverence. We must do the same. We need to feel an emotional attachment, maybe even a spiritual attachment, to the land. We need to nurture the land and its resources, not exploit them. The land will reward us for doing so, for if the land is healthier our farming will be more productive and sustainable.

The truth is that most Australian farmers do have an emotional attachment to the land. They might not admit this, of course, but I have no doubt it is true. It is the reason they are farmers. What other reason could a farmer have for working 50 or 60 hours a week for not much money? Of course, it might be said that farming is a lifestyle choice, that farmers choose to be farmers because they like the lifestyle. In my opinion—and I speak as someone who has been a farmer for more than half a century—one of the main reasons they like the lifestyle is that, in their own way, they love the land. So the idea of farmers attaining an Aboriginal-like reverence for the land is not really so revolutionary. They are already halfway there.

One of the biggest frustrations of trying to explain—or justify—why I have invested so much time, effort and money in protecting and rehabilitating the environment at Lanark is that many people assume I am a head-in-the-clouds greenie and always have been. This is why I am quick to tell people how badly I used to treat the land when I first took over the property—how I did not hesitate to use chemicals and how I allowed it to be severely over-grazed. I want them to know that I have done the other thing—that I know the ins and outs of it from personal experience and I know the debilitating effect it has on the land. Maybe you could get away with flogging the land in, say, New Zealand. You cannot get away with flogging it in Australia. The ecosystem here is just too fragile.

This book is not really my story or Cicely's story. It is the story of Lanark, a 1715-acre block of land in Victoria's Western District that was all but dead and was brought back to life. I hope the story may help others who want to return life to their own land.

2

A BOLD EXPERIMENT

Looking back, I realise now that Lanark was essentially an experimental farm. Most of the work that Cicely and I did there—the tree planting and other environmental projects—was experimental in the sense that we did not know for sure what the outcome would be. No other farmer that I knew of had done what we set out to do, so there was no past experience for us to rely on. Cicely and I were flying blind. We could not be certain that the huge investment we were making in the environment at Lanark would not prove as big a waste of time, effort and money as all the sceptics around us seemed to believe it would be.

Lanark was not an ordinary experimental farm, however. Having made a mistake, perhaps by planting the wrong trees in the wrong place, we did not simply remove the mistake and start again. All our mistakes—wrong species of trees, badly located fences and so on—are there today. In this sense Lanark has always been a work in progress. If you were to drive around the property today, you could trace its entire history over the past 50 years, for it is all still there, the failures as well as the successes.

So what exactly do you see if you drive around Lanark today? How is it different from other Australian sheep properties? Well, the thing you would probably notice first is that there are a lot of trees. I estimate that we have planted 100 000 of them over the years, and many others have grown naturally through regeneration from my early plantings. About 30 per cent of the trees, occupying an area of about 75 acres (30 hectares), have been grown for timber and will eventually be harvested. There are another 35 acres (fourteen hectares) of shelter belts, consisting of various native tree species, both indigenous and exotic. The rest of the trees, all native species and mainly indigenous, are in wildlife habitat reserves totalling about 90 acres (36 hectares). In all, about 200 acres (80 hectares) of land at Lanark, representing just over 10 per cent of the property, are covered by trees.

Then there are the reinstated wetlands. All in all, they occupy almost 120 acres—or around 7 per cent of the property. There is a seventeen-acre (seven-hectare) lake, named Lake Cicely; about 30 acres (twelve hectares) of seasonal water meadows; and 70 acres (28 hectares) of swamps. The rest of Lanark, amounting to just over 75 per cent of the property, consists of grazing land.

I said in the previous chapter that what we have done at Lanark has been good both for the farm and the environment. Now, it is not hard to see why reinstated wetlands and tens of thousands of trees forming bushland reserves would be good for the environment, including, in particular, bird life; but it has also been good for the farm in a number of important ways, which I will discuss at greater length later in the book.

Trees shelter the stock from the worst of winter weather, which, among other things, means that the stock do not need to eat as much grass to stay warm. If they are aligned in windbreak formation, trees are able to shelter crops and pastures from the wind, thereby reducing evaporation from the soil in summer and generally encouraging growth. Trees create fertility by raising minerals to the surface and by simply dropping their leaves and branches. They can eventually be harvested as a cash crop if they are planted and managed as timber trees.

Trees also raise the capital value of any farm where they are planted. In 1983 I arranged for the Victorian Farmers' and Graziers' Association (VFGA) (later to become the Victorian Farmers Federation) to conduct a study of land values at Lanark. The aim was to compare the value of land where there had been an intensive planting of trees and a restoration of wetlands with the value of land that was bare. The valuations were done by a registered valuer. I had the same portions valued again five years later. The value of the land with trees and shelter was then 78 per cent higher than the value of the land with none. What I found even more interesting is that during that five-year period—a period in which, of course, the trees grew much bigger—the value of the land with shelter and associated wetlands rose by 119 per cent, whereas the value of the land without shelter rose by only 29 per cent. The study appeared in *Financial Benefits of Farm Trees*, published by the VFGA and the Victorian Garden State Committee.

One other factor tends to be overlooked, yet I consider it enormously important: quality of life. I am referring here to the quality of life of the farmer. Living and working on a farm with plenty of trees (with the shade they provide) and shrubs is far more satisfying than living and working where it is bare. I know this for a fact, because I have experienced both.

Prue Powell expressed this same view in an article she wrote about Lanark for the book *Bushland on Farms: Do You Have a Choice?*, which Rob and Steve Davidson published in 1992. 'As the quality of the land rises through better management and growth of remnant vegetation and specifically placed blocks and belts of trees', Prue Powell wrote, 'so does the quality of life for the people working in that environment. For example, there is a stark contrast between working in exposed treeless yards with sheep or cattle and working in yards protected and shaded by a plantation of trees and shrubs. I know which I would opt for!'

3

IN THE BEGINNING

My family has owned Lanark since 1937. Every story has its beginning, and I suppose Lanark's story really begins with the first Fenton to arrive in Australia—my great-grandfather, David. Fenton is an old English name, but there are many Fentons in Scotland, too, and my own family is Scottish by origin. Our branch of the Fentons came from a village not far from Edinburgh, where a lot of the Fentons were concentrated. A short drive east of Edinburgh there is a fortified tower—a small castle, really—which dates to the 1500s and is called Fenton Tower. In recent years it has been turned into up-market accommodation for tourists. I have only seen photos, but what I find interesting is that there is not a tree in sight. It has occurred to me that if my own Fentons came from there originally they may well have arrived in Australia feeling instinctively uncomfortable with trees around them.

David Fenton came to Australia with his wife, Mary, in 1852. Gold had been discovered in Australia the year before, and we may assume it was gold that enticed David to Australia. By 1856, he and Mary had settled with their young family at Victoria's premier gold town, Ballarat. For a while he scratched around for gold in the Ballarat district, apparently without much success, and later ended up at Ararat, a handsome town about 100 kilometres west of Ballarat. In 1862, just ten years after arriving in Australia, David died of tuberculosis in Melbourne. He had apparently been separated from Mary for some time and had drifted to Melbourne in search of work. From what has been passed down about him, it seems he was a heavy drinker and, accordingly, had neither money nor the ambition to earn it.

At the time of her husband's death, Mary was 31 years old and still living in Ararat with her children, one of whom was my grandfather, John Fenton. A few years later she married an Ararat businessman named James Donisthorpe Smith, who, among various interests, had a grocery store and a wine-and-spirit business. By all accounts, he treated his stepson John quite well, and for a time paid for him to attend a private boarding school at Geelong. John left school at around the age of fifteen and began working as a delivery boy for his stepfather's grocery business. He rode around Ararat on a bicycle with a basket at the front, delivering goods to the store's customers. The story goes that one day, when John was about sixteen, he overheard a couple of people talking about a property that was about to go on sale in the town. He gleaned from the conversation that the property would be a very good buy and, apparently by alerting his stepfather to the opportunity, was able to make some money from the transaction.

For young John, it was to prove the first of many successful deals in real estate, a business that he later made his career.

Before that, though, he took another step that was to shape the rest of his life: he married a girl named Helen Laidlaw, known to all as Nellie, who belonged to a family well established in the Amphitheatre district. (Amphitheatre, a name that may sound odd to motorists from other states who happen to drive through, is a small township on the Pyrenees Highway between Ararat and Maryborough. The township was originally located in a natural amphitheatre, from which it took its name, but was later moved to another site on the highway.) John and Nellie were married in 1884 and made their home in Hamilton.

John Fenton became a stock and station agent. He ended up with his own agency, to which he admitted several partners, including his son-in-law Bill Beggs, of whom more will be said later. But John Fenton retained control of the business until it was sold to Australian Estates in 1937, after which he retired at the age of 81. He died in 1940, aged 83.

Thanks to the success of his business, John Fenton became quite wealthy. He came to own, either individually or in partnership with others, numerous rural properties as far away as Forbes in New South Wales. For a couple of years, from 1904 to 1906, he was Hamilton's mayor. He was an interesting character: reserved and subdued in manner, yet kind in his own quiet way. He was rather serious and inflexible, and it seems that people meeting him for the first time found him a little intimidating. He was also a teetotaller. The tragedy of his life was the loss of one of his two sons, Jack, in World War I, just five months before the war ended. Born in 1892, Jack was older than his brother Dave, my father, by a couple of years. He was also a little taller than Dave and better looking. He was better at sport than my father, too. All in all, he was the apple of everyone's eye—and he was certainly the apple of his father's eye.

Jack was 26 when he was killed in France. When news came of Jack's death, my grandfather was inconsolable. Among our family relics is a photo of a family gathering in Hamilton to farewell Jack before he went off to war. It was obviously a dinner, for the people in the photo are dressed up to the nines, and with one exception everyone looks sad and despondent about Jack's impending departure. The exception is Jack himself, who was obviously pleased to be going. Jack was one of 79 men from Hamilton killed in World War I, a dreadful toll which affected the town in many ways. For instance, my mother's oldest sister, Berta, who was born in 1902, did not marry, and nor did many other Hamilton girls of her generation. The reason so many women of the same generation never married was, of course, that the men they might have married were dead, killed in the war. Aunt Berta had a large group of unmarried female friends of about her age in Hamilton, many of whom were still getting about in the 1980s. (Berta herself lived until 1998.)

I have been told that my grandfather's attitude towards my father changed after Jack died, and he became sterner and more rigid than before. I am not sure of the reason, but I suspect he may have felt, subconsciously, a tinge of resentment at the fact that David was alive while Jack, almost certainly the favourite, was dead. It was not as if my father shirked military service, incidentally. He joined the Australian Flying Corps and was due to go overseas, but, fortunately for him, the war ended before he left.

After the war, probably in 1919 or 1920, my father joined his father in the stock and station business as a junior partner. Not long after this my grandfather posted a notice on an office wall that was clearly directed at my father. It advised everyone working there that the office car had to be garaged at 7 p.m. each evening and was to remain garaged until 7 a.m. the next morning. It seems my father had been using the car for fun after working hours, and my grandfather decided to put a stop to it. My father was never invited to take over the business, and he was still a junior partner when my grandfather eventually sold it.

My father may not have been as good a sportsman as his brother Jack, but he was still a fine tennis player and good enough at cricket to represent a Hamilton district side against the English team that toured Australia in the summer of 1920–21. The match was played at Melville Oval in Hamilton. David Fenton made 17 not out in the first innings, his team's second-highest score, but in the second innings was bowled for a duck by the great England opening batsman Jack Hobbs.

As it happens, Hobbs's opening partner for much of the 1920s, Herbert Sutcliffe, visited Hamilton too, although on a separate occasion. There are a number of photos of him at my grandparents' home, where he stayed. According to family legend, Sutcliffe took a liking to my father's sister Nell, so much so that they 'stepped out' together during his visit. I may add here that Nell Fenton never married. She spent her life caring for her father, John.

Although my father never became a full partner in the family firm, he did well enough to be able to buy Lanark when it came on the market. It seems he handled much of the sheep and cattle side of the business, while my grandfather concentrated more on valuing and selling properties. One of the properties my grandfather bought before World War I was a 9725-acre (3900-hectare) property near Cavendish, north of Hamilton, which he earmarked for Jack. The property never came to

Jack, of course, and it did not come to my father, either. In his will, old John Fenton left it to his four daughters. This was, I suspect, another indication of how John's attitude towards my father was soured by the loss of his other son.

4

A 1715-ACRE TRIANGLE

In 1937 a property about 25 kilometres southwest of Hamilton, called Bassett Estate, came on the market. It was named for Christopher Bassett, who had taken up land in the district a century earlier but had been murdered there by Aborigines in 1843. In 1937 Bassett Estate was close to 10 000 acres (4000 hectares)—big but by no means the biggest in this part of Victoria. The property was closely associated with the adjacent village of Branxholme; each was dependent on the other. Most of the shearers and other people who worked there lived in the village, and Bassett cattle were always loaded at the Branxholme railway station yards.

I do not have a record of the auction, but it was almost certainly held in the old Victoria Hotel, then one of Hamilton's landmarks—a three-storey, bluestone building of character, which had once been a Cobb and Co station. (Sadly, it was demolished about 40 years ago and replaced by a much inferior commercial building.) Bassett was an excellent property in every respect, with very productive river flats, so there was plenty of interest. My grandfather certainly had his eye on it. He did not want it for himself but for one of his daughters, Jean, who was married to Marcus Whiting, the son of a Melbourne businessman. The Whitings were well off. After their wedding, the couple were able to spend nine months honeymooning in Europe. (They are said to have gone to the theatre in London no fewer than 27 times.) They already owned quite a large dairy farm near Warrnambool, where they must have lived in some style, for they employed several domestic staff, including a housemaid. But it seems my Aunty Jean was keen to buy a property somewhere in the Hamilton district, which for her, of course, was home territory.

At the auction, a group of people, mainly local farmers, bid jointly for the property, the plan being that each would end up with a designated section of it. The Whitings were part of this group, and they emerged with a 4000-acre (1600-hectare), rectangular slab of Bassett Estate that included the homestead and retained the name Bassett. My father was part of the group, too, and he ended up with a 1715-acre triangle in the middle. My father was helped at the auction by his friend Eric Reid, a stock and station auctioneer. After the auction, he and my father retired to the bar of the Victoria Hotel to celebrate the purchase. While they were having a drink, Reid asked my father what he intended naming the property, and my father replied that he had no idea. Reid said, 'Your family came from Scotland, didn't they? You ought to give it a Scottish name.'

So they asked the publican if he had an atlas, which he did. The atlas was placed on the bar counter, and Reid and my father pored over a map of Scotland. They saw the name Hamilton and looked for a suitable place name nearby. Reid put his finger on Lanark, a small town southeast of Hamilton. 'It's about the same distance from that Hamilton as your property is from this Hamilton,' he said. 'So why not name it Lanark?' My father thought this was a good idea, so Lanark became the property's name.

Marcus Whiting died, aged 57, just two and a half years after buying Bassett. He had been supervising the loading of a mob of cattle at the Branxholme railway yards, and he suffered a fatal heart attack as he went to mount his horse for the ride back to Bassett. His overseer, Alf Turner, ran to the railway station to get help, but it was too late. My father stepped in to help Aunty Jean run Bassett. He was then responsible for several other family properties in the district, including one named Tahara South, which belonged to his father. This is not to say he was in any way a hands-on manager. We lived in town, and he employed men to run the various properties on a day-to-day basis. I have only one boyhood memory of going with him to visit Lanark—and he was there that day not to check on the sheep but to shoot ducks. I can picture him now: sitting with a mate on a couple of wool bales, waiting for wild ducks to fly in to the remnant of a swamp that existed there then.

My parents had been married at St John's Church in Toorak, Melbourne, in 1931. My mother, Lesley Baker, was a town girl, daughter of a solicitor, and she grew up in Hamilton itself. She was a third-generation Australian. Her maternal grandfather, William Renfrey, was one of many Cornish immigrants who settled in South Australia's so-called Copper Triangle, an important copper-mining area in the late 1800s and early 1900s. His occupation was building chimneys for the smelters there, although he later moved his family to Hamilton. My mother had two sisters but only one brother—my Uncle Vern, of whom I will have more to say later. She was young, good-looking and popular, so at the time she married my father, who was then aged 33, she was considered quite a catch. My parents had two children, a small family in those days. I was born on 25 February 1935 and my sister, Robin, three years later. We were delivered by a midwife, Sister Marjorie Brook, who later served with distinction in World War II. She became a family friend, and I remember her with affection.

I was much too young to be aware of this myself, but it seems that by the 1940s my father had begun to suffer poor health. He was well enough to go off fighting the bushfires of 1944, which did so much damage in Victoria (32 people died and 700 homes were destroyed), but a few months later, in April 1944, he died, aged just 46. No fewer than 170 wreaths were sent to his graveside, reflecting his prominence in Hamilton. He had been a Hamilton town councillor and a president of the Hamilton Club.

I was nine years old when my father died, and my mother was in her early thirties. A widow as young and attractive as my mother might well have married again, but my mother never did, although she was to live until 1981. One reason, I suspect, was that my father's sisters, out of sympathy, took 'poor Lesley', as I later heard them refer to her, firmly under their wing. They were kind people and they meant well, but I am sure they restricted my mother's social life.

It must have been terribly difficult for her, but my mother kept our family life going as best she could. We still went to Portland—a coastal town about 90 kilometres south of Hamilton—for our annual holiday, for instance. My father had always rented a house for us there. It was common then for Portland people to let their homes during the holiday season and move into a shed at the back of the garden. My mother kept the tradition going. Each year we headed off to 'Porty' for a holiday and the first one to see the sea was rewarded with a penny ice cream.

Over the years I have met numerous people who, like me, had the misfortune to lose a parent while they were young children. Some I have spoken with have very clear memories of the parent, even though they were perhaps only four or five years old when the parent died. Not me. My memories of my father are few and fragmented. Here is one of them. When I was eight and Robin about four, our father planted trees for both of us in the garden of our home at 20 Chaucer Street, Hamilton. We helped him plant them. My tree was a claret ash and Robin's a silver birch. The claret ash grew just outside the closed-in verandah where I used to sleep, so I was able to keep track of its progress.

Then, a few years after my father died, my mother had the claret ash chopped down. It had grown into a sizeable tree by then, and my guess is that its roots were judged to be a threat to the foundations of our double-brick home. Whatever my mother's reason was for getting rid of the tree, my reaction was extraordinary. By all accounts I ranted and howled for two or three days afterwards. Nothing could console or placate me. Even now, I can recall the sense of outrage I felt. Something precious had been taken from me and I would never get it back.

My reaction to that tree's removal was certainly out of character. I cannot remember another occasion, before or afterwards, when I put on a turn like that. Without wishing to sound overly introspective so long after the event, I am sure now that I must have somehow associated the tree with my father. Subconsciously, I may have considered it my last personal link with him.

That strange episode may also show that, even then, I was already starting to develop a fondness for trees. Something else

I remember from that period may have been a sign of things to come. A few of the boys I knocked about with in Hamilton had armed themselves with home-made shanghais—or gings, as we preferred to call them—which they liked to fire at birds. We would ride out on our bicycles to the town's Lutheran Cemetery, where there was some natural bushland, and the others would fire their gings at the blue wrens, silvereyes and other native birds they found there. I went along with them, but I disliked the idea of shooting at birds and never did it myself. Did this suggest I was already developing an environmental conscience? Perhaps, although it might also mean I simply had no stomach for blood sports.

By now I had certainly started to take an interest in gardening. One of our near neighbours was an elderly gardener named Russell Pittman, of whom I have fond memories. He belonged to a class of Australian pioneers who have never received the recognition they deserve. When we think of early pioneers we tend to think of the people who first settled the land, cleared it and farmed it. But there were pioneers in country towns, too—those solid working-class people whose skills and labour kept the entire rural community operating.

Russell Pittman was one of these. He lived in a small cottage two streets away from us, and he used to come to our house and show me how to plant vegetables and other garden plants. I have a clear memory of him sitting on our side verandah, having a cup of tea with my mother and eating the scones that she made specially for him. That was characteristic of people then: they were not constantly on the go, as people tend to be today, and perhaps for that reason they did not earn as much. But they did have time to spend with other people. Not many workmen have time to sit and yarn over a cup of tea today.

Under Russell Pittman's direction, I worked up a garden just outside my sleep-out. My Aunt Nell used to give me nasturtiums to plant there. It was a wise choice on her part, because nasturtiums grow easily, which makes them an ideal plant for the child gardener. I was eight or nine years old then. Was my interest in gardening as a child a precursor to my passion for planting trees later in life? Probably, but I do not know for sure.

After my father's death, my mother came to depend financially more and more on the wool-growing operation at Lanark. Under the terms of my father's will, the property was controlled by the Ballarat Trustee Company. Indeed, a sizeable proportion of the properties in the Western District that had been left in a will seemed to be controlled by the Ballarat Trustee Company, which later became the Union Fidelity Trustees Company. Engaging a trustee company in this way was part of a simple device to avoid death duties. Instead of leaving your property to your spouse, who would have to pay death duties again when he or she died, you would leave it to your children or, as often happened, to your grandchildren. This was called skipping a generation.

This is what my father did. When my sister, Robin, and I were still small, my father left Lanark to our future children with the proviso that my mother would have the income from the property during her lifetime. Until our children were old enough to take over the property, it would be controlled by the trustee company. This was a fairly common arrangement. The aim was commendable—avoiding two generations of death duties—but it was not such a good idea in practice. The trustee company was the problem. For one thing, it charged substantial fees. The fact that the trustee company controlled the property was a terrible bind, too. We could not do what we wanted to do with our own land. This would have been more tolerable if the trustee company had run the place efficiently. In fact, many of the decisions the company made about what should and should not be done at Lanark were, in my opinion, hopelessly wrong.

As it happened, my mother had no affinity with Lanark and certainly had no idea whatsoever of how to run a sheep property. We lived in Hamilton itself—and Lanark had been managed for some years by an old bachelor named Donny Macdonald, who lived there in a hut. That hut, a weatherboard lean-to, was the only structure on the property, for at that time Lanark did not even have sheep yards. The hut is still there today and, for me, it always brings back memories of Donny. He was a small man, tiny really, with a subservient manner. His hut consisted of three rooms. One was a workshop about one metre by three; another was a feed room about the same size, which was used for storing chaff for Donny's horses; and the third was Donny's living quarters—a room about three metres by three, which contained a stretcher bed, a few cupboards and an old wood stove with an iron chimney. The most modern thing in the room was a telephone, which had a single direct line to the Bassett homestead next door.

Donny Macdonald got around the property not on horseback but in a horse-drawn cart. In those days Lanark generally carried only 1000 or so wethers at any one time, so for most of the year Donny would not have had a lot of work to do. I imagine he mustered the sheep no more than twice a year, once at shearing time. There was no shearing shed on Lanark, so Donny would have taken the sheep to the shed at Bassett, about a kilometre away. For the rest of the year, he would have mended fences as required, made sure the stock had water, and generally kept an eye on things. He would certainly not have done any spraying or fertilising. Donny owned a bicycle, which he rode into Branxholme each Friday, a three-kilometre journey, to get his supplies. From week to week, it was probably the only human contact he had.

I am sure Donny had spent his whole life living and working on sheep stations. He may even have begun his working life as a shepherd, for he was probably just old enough to have done that. Until the late 1800s, when there were vast, unfenced sheep runs, graziers employed men to travel about with the sheep for weeks on end. Their job was to keep the sheep together in a mob, make sure they had feed and water, and protect them from dingoes. They were called 'shepherds', although they had little in common with shepherds as they were known in Europe. They led a very solitary life, as Donny did until he died. One day in the mid-1940s, my Aunty Jean at Bassett took a call from Donny on the direct phone line. He said to her, 'Mrs Whiting, I'm very sick.' She asked him how sick. He said, 'I think I might die.' This sounded serious, so Aunty Jean immediately drove in her car to Lanark. Donny had not been exaggerating: he was lying in his stretcher bed, dead.

As I said before, Donny's hut is still standing. About six years ago I happened to walk past the hut with my grandson Will, who was then about seven years old. He remarked on the fact that the hut looked a wreck, to which I replied that I hoped he would help me restore it. Will then said, 'Well, we'd better do it quick. My dad says he's going to knock it over with his tractor and burn it.' Will's dad is my son David. I gave Will a brief lecture on the need to preserve the heritage of the countryside, and I later prevailed upon David to spare the hut, which he has.

After Donny died, the man my Aunty Jean employed as a manager at Bassett, Alf Turner (always 'Mr Turner' to me), also took over the day-to-day running of Lanark. I know a little about Turner, because he talked about himself while we were working together in the 1950s and I recorded some of the details in my farm journal. He had started work at Bassett in 1922. He arrived there by train in the company of 11 000 wethers that had been brought down from Swan Hill.

In the 1940s, not long after the war ended, the trustee company approved the construction at Lanark of a shearing shed and a proper house to accommodate a manager. The house was built for £1400 by Athol 'Attie' Gunn, the father of my friends Murray, Graeme and Noel. It was occupied first by a returned soldier, Don Macgugan, who was not there long because he was soon allocated a soldier-settlement block of his own. The next manager to move in was an ex-shearer named Keith Fort, who knew all there was to know about farming. Fort was still Lanark's manager when I took over the place in the late 1950s. After Cicely and I were married, by which time Fort had left, the manager's house became our home.

5

SLOW BUT RELIABLE

In 1945, the year after my father died, I came down with a serious ailment that was to leave its mark on me for life. It was an acute abscess in a frontal sinus—that is, the sinus in the bone above the eye. In my case, it was the left frontal sinus. The first symptom I had was a bad headache. My mother took me to a GP in Hamilton, who could find nothing wrong with me. In fact, he warned my mother against pandering to me because I no longer had a father. I was then ten years old. A day or two later I collapsed and was taken to the local hospital, where the cause of the problem was identified, probably by an X-ray. My condition must have been considered serious, because a Melbourne surgeon named Scantlebury was flown to Hamilton specially to treat me. What he did was to cut through the bone above my eye to drain the abscess. My mother told me long afterwards that during the operation Dr Scantlebury emerged from the operating theatre to ask her whether she wanted him to continue. He was warning her, in effect, that I might not survive the operation, and he wanted her approval to keep going. She told him to press on, which, fortunately, he did, for the operation proved successful.

I was in hospital—the old McKellar Hospital in Hamilton, which is now part of the Hamilton Hospital complex—for a long time afterwards. There was an RAAF airman in the same ward, a young man from a grazing family in the district, who had been wounded in the war. He taught me to play euchre. He had a pack of cards, and I would sit on his bed and play euchre with him for hours. I did not realise it at the time, but the airman was dying from his injury. I heard later that he died not long after I left hospital and went home. I have thought about him many times since. I like to imagine I may have been some comfort to him—a kid sitting on his bed playing cards. He certainly helped my own recovery. Eventually, I did recover completely, although the operation left a deep scar above the eye which I bear to this day.

In the late 1940s I began spending a few days at a time with Aunty Jean at Bassett. Sometimes I brought along a school mate of mine, Roger Harrison. We went rabbiting with Alf Turner's son, who transported us about in a horse and cart. We also sailed about the property's deepest dam on a raft made from oil drums. We did it without life jackets, of course. I was only twelve or thirteen then, yet Aunty Jean used to say to me, 'You know, dear, that you're going to have to run your farm when you grow up.' The idea did not appeal. I knew nothing about farming and, in any case, I was happy living in a comfortable house in Hamilton. Those visits to Bassett were, however, my first real experience of farming life.

In 1949 my mother sent me off to board at my father's old school, Geelong College. Before that, I had gone to school at Hamilton College, which was convenient for me, because we lived close to its front gate. In fact, no student at Hamilton College lived closer to the school than I did. At Geelong College, I proved to be an undistinguished student, as my 1949 school report illustrates. I came thirty-first in a class of 35 and was close to last in most subjects. The school principal wrote of me in the report: 'A reliable, trustworthy boy but slow academically.' A few 'special comments' were added about my poor marks. One of them, in reference to English, said, 'Grammar weak.' Another, in reference to Latin, said, 'Subject seems beyond him.' The final summary, presumably by my teacher, was: 'Does some good work sometimes, but is very unstable. Not sufficiently interested in accurate thinking.' Things did not improve in the next few years, either. I failed Year 10 and had to repeat it, after which I left school for good. That was in 1952.

During my time at Geelong College—in the summer of 1950–51, to be precise—the school had the fairly rare distinction of having two old boys in the Australian cricket team at one time. One was Lindsay Hassett, the Australian captain, and the other a spin bowler with a unique 'flicking' method of spinning the ball, Jack Iverson, who was not quite two years younger than Hassett. I have read that Iverson once bowled Hassett in an inter-house match at the school, which was no doubt a sign of things to come. In 1952 Hassett returned to the school to play for an old boys' team against our first XI. I was the official scorer that day, which was probably the high point of my cricket career: cricket was not my game. I really enjoyed playing footy, by which I mean Australian football, but the sport I did best at—and enjoyed most—was athletics, for I could run fast enough to make the relay team.

Geelong College had a large boarding house in my day, which was occupied mainly by the sons of sheep graziers from various parts of Victoria, especially the Western District, and from the Riverina. Many of these boys intended going on the land, and for some of them it was almost an automatic thing to go from school to an agricultural college. This is exactly what I did. The decision was not mine, however. It was made by my mother on the advice of Dick Hughes, the manager of the Union Fidelity Trustees Company, who had been a close friend of my father's. Dick Hughes was a fine, upstanding character who gave my mother a lot of support while we were growing up, although, as I later discovered, he did not have much practical knowledge of farming. He was probably right, though, to suggest that an agricultural college would be good for me.

I enrolled in Longerenong Agricultural College, which is just north of the Grampians, quite close to Horsham. Because I had passed only the intermediate exam at school I went into first year at the college. Boys who had matriculated went into second year. This put me at a social disadvantage, since I was seventeen going on eighteen and sharing the classroom with boys of sixteen. But I made some good friends there and occupied my spare time happily by playing footy and other sports.

Longerenong was a great training ground: we acquired practical skills there as well as theoretical knowledge. We might spend one week in a dairy, the next in a poultry shed, the next in a woolshed or abattoir or even a bakery. I hated the abattoir work. First-year students like me had the job of slitting the sheep's throats, but I freaked out whenever it came my turn to do it. (I did get over the revulsion eventually and for a long time I killed our own meat at Lanark. But about twenty years ago the old aversion to slaughtering animals suddenly returned and I was never able to kill our own meat again. Today, I would find it hard to kill a fly.)

The one thing that was lacking in Longerenong's syllabus was some reference to the importance of planting trees on farms. In all the classes I attended there, as best as I can recall, the subject was not mentioned once.

I enjoyed Longerenong and learned quite a lot, but unfortunately my time at the college was cut short. Here is what happened. Someone who had been in our football team the year before, but who had since left the college, decided to revisit his old team-mates. On his way to the college he called at a local pub and bought a few long-neck bottles of beer, which he shared with us when he got to the college. It was both a generous and unwise thing to do, for at that time Longerenong had a total ban on alcohol: anyone caught drinking was liable to be expelled. Since our old team-mate had been kind enough to bring the beer, however, we felt obliged to drink it. I had one glass. Others may have had two glasses, but that was all. When we went into the dining room for dinner that evening, we were stopped at the door by the head prefect of the college, who apparently could smell the beer on our breath. We were apprehended and made to sit at a separate table in the dining room. There was much worse to come next day. Our house master ordered us to pack our things and leave Longerenong immediately, which we did. We were taken to the railway station at Horsham in the back of the college's farm truck.

Six of us, all members of the football team, were later formally expelled. Some of the parents reacted with outrage and complained so loudly that the story of our expulsion eventually made page one of the Melbourne *Sun*. It was all acutely embarrassing for the college, which then sought to defuse the situation by inviting the six of us to return. The house master who booted us out was reportedly made to resign. I felt sorry for him even then. He was close to retiring age and had obviously felt he was just enforcing the college rules. The catch was that we could not come back that year: we had to wait until the following year and then start the academic year again. I was already nineteen by this stage, and I decided I was too old to start second year at the college again.

I decided to go jackarooing.

6

A RACE AGAINST HECTOR HOGAN

My first jackarooing job was at a historic property in central Victoria that was owned and run by a hard-bitten old character, a prominent grazier in his day, whom I will call Mr White. This was not his real name, but, given the memories I have of him, it might be better not to identify him or his property. At the time I went to work for him, White was still breaking in a lot of rough hill country. Among other graziers, he had a reputation for being a shrewd, skilled, extremely competent operator.

I was quartered in a stone outbuilding just behind the main homestead. On my first morning there I woke to hear a large pack of dogs barking. A moment later there was a triple whip crack. I looked out the window to see old White in his nightgown, cracking his whip to stop the dogs barking. It turned out that the dogs belonged to a rabbiter who was employed full-time on the property, which, like many other properties in that area, was overrun with rabbits. There must have been as many as 30 dogs there, all crowded into a big stock crate. I never saw the dogs in action, although I assume their job was to catch and kill any rabbits that escaped after they were flushed out of their burrows into a net by the rabbiter's ferrets.

That same day White assigned me my first job: milking the house cow. I had to do it by hand, of course, and I was not much good at it, which he was quick to notice. A few days later he came to me and told me to follow him. He walked down to the stables, with me behind, and when we arrived there he said to me, 'Saddle my horse.' I replied, 'I don't know how to, Mr White.' For a moment he looked stunned. Then, he exploded: 'What! Don't know how to saddle a horse!' He was so annoyed he even threw his hat on the ground. So he saddled the horse himself, and he also saddled the horse I was supposed to ride. 'Get on,' he said. I then had to admit to him that I could not ride. This sent him into a rage that he struggled to contain. 'Very well,' he said in a threatening voice, and he unsaddled my horse and put it back in the stable. He then mounted his own horse and, as he rode off, said, 'Follow me.' This is what I had to do: follow him on foot as he jog-trotted his horse right around the boundary fence, a distance of maybe fifteen kilometres.

White was by now convinced I was entirely useless. Whatever the truth of that, the fact was I was terrified of him, which hardly made for a good boss–jackaroo relationship. So I was probably asking for trouble when I mentioned to him some days later that I would like to play football in the nearby town each Saturday. I did hasten to add that it would not affect my work on the property in any way. 'Not a hope,' White told me. 'You're not playing football.' I persisted, saying I had always played football and that, in any case, it would be good exercise and would improve my fitness. 'If you need exercise,'

White said, 'you can get plenty of it here. Picking up stones.' He then gave me the job of trudging behind a bulldozer that was breaking in paddocks on the property and picking up the stones that the bulldozer brought to the surface. Having picked them up, I had to pile them in a heap, where they could be collected later by a front-end loader.

Picking up stones became my main job from Monday to Friday. Then on Saturday I had to drive the bulldozer in place of the regular dozer driver, who had weekends off. Sunday was my one day of rest. I had never driven a bulldozer before, but I had already learned the basics of it from the dozer driver between picking up stones. I think he felt sorry for me, and during breaks from work he used to show me how the dozer worked. I soon got the hang of it. In fact, I thought that driving the bulldozer—pulling the various handles and being in command of such a powerful machine—was great fun. Unfortunately, I had an accident in the dozer that brought White's wrath down on my head again. While clearing land close to a creek, I slid the bulldozer a little way down the creek bank and clipped a tree with the discs attached to the back. There was not much of an impact, but it was enough to bend the axle of the discs. White was furious with me, yet to my surprise he did not ban me from driving the dozer. It remained my Saturday job, so playing football was still out of the question. In fact, driving the dozer proved to be a valuable skill, which was often useful to me in later life.

Like all jackaroos, I was paid a measly wage, but since I had nothing to spend my money on I was able to save pretty well everything I earned. Within a few months I had saved enough to buy an old Chevrolet ute from the rabbiter who worked on the property. It cost me £40. The vehicle was actually a Chevrolet car that had been converted into a ute. It did not have a cabin as such—just a kind of hut made of flat iron nailed to a wooden frame. The Chev also had no lights and no brakes other than a handbrake, yet it ran well. About this time White and his wife decided to go overseas for a holiday. By now, I had long wished to resign and leave the property, but I was too frightened of old White to tell him. His departure with his wife made things easy. The day after they set off I went to the property's manager and told him I was leaving. He said, 'You can't do that. The boss has just gone—you can't leave now.' I replied that I didn't care whether the boss had left or when he would be back and I didn't care whether or not I was paid the money owed to me—I was going.

I did go, too. I packed my few possessions and drove the Chev down the Hume Highway to Geelong, which my uncle and aunt, Vern and Gladys Baker, had recently made their home. Uncle Vern and Aunty Glad had sold their sheep property near Hamilton and moved to Geelong so they could send their sons as day boys to Geelong College and their daughter, Jill, as a day girl to the Hermitage, a girls' school in Geelong. I drove all the way to Geelong without brakes but, happily, also without incident. On arriving at the Bakers' house in Stephen Street, however, I saw they had an empty garage. I swung the Chev into the garage, intending to park it there, but the handbrake would not stop the car quickly enough and I ploughed through the wall at the end of the garage. It was an embarrassing start to my new life in the Geelong area, which at the time I put down to bad luck. I now think it was remarkably good luck that I travelled safely, without brakes, along the Hume Highway as far as I did. I may add that the Chev never ran again. Uncle Vern towed it to Point Wilson, his sheep property at Lara near Geelong, and used the motor to drive a mechanical sieve.

Uncle Vern gave me a job as a farm hand on Point Wilson. Each week he and I bached together on the property (by which I mean we lived as bachelors, doing our own cooking, washing and cleaning) and at weekends we returned to the family home in Geelong, which enabled me to play footy for Lara, one of the teams in the Geelong Football League. Suddenly, life felt good again. Mr White and his property had already become a memory.

Uncle Vern had two other farm hands working on the place. One was a huge man who was known, predictably, as Tiny. He and his wife had rooms of their own in the property's old stone farmhouse. Before coming to work for Uncle Vern, Tiny had been a rigger in a circus. The other farm hand was an old chap named Joe, who was probably an alcoholic. Invariably, his supply of grog would run out by the middle of the week, and by Thursday he would start asking Uncle Vern, over and over: 'When are we leaving tomorrow? What time do you think we'll get away?' In fact, we normally left for Geelong around three o'clock on Friday afternoon, and Joe was the first in the car. As soon as we reached a pub on the outskirts of Geelong, he would say, 'Thanks, Bern—this'll do,' and get out. He called Vern 'Bern' because he could not pronounce V. Uncle Vern would pick him up on the way back to the farm on Monday morning.

We four were an unusual cross-section of people, I suppose, but we blended well. We all lived rough on the farm, including Uncle Vern, who did most of the cooking for us. Periodically he would kill a sheep for meat, which he cooked on an open fire, and he liked making mutton soup by the gallon in an old kerosene tin. Every scrap of the sheep's carcass was consumed in the end. Nothing was wasted, not even the sheep's head, a frugality that had probably been acquired during the Depression.

In the winter, Uncle Vern would often have a sheep's head rotating in an old kerosene tin filled with boiling water. Every so often, as the sheep's head rotated, one of its eyes would stare at me, which I found unsettling.

At Point Wilson I slept in a stone room out the back. On my first night there I was kept awake by rats scuttling about in the roof. Next day Uncle Vern gave me a shotgun and a supply of rat shot. 'When the rats start making a noise', he told me, 'just fire at the noise through the ceiling.' I did manage to shoot a few rats in the weeks that followed, although the ceiling ended up looking a little tattered. Another thing I remember about Uncle Vern: he had enormous faith in eucalyptus oil as a cure for all types of ailments. His personal favourite was blue mallee oil sold under the old Bosisto brand, which took its name from Joseph Bosisto, the man who began the eucalyptus oil industry in Australia back in the 1850s. The Bosisto company used to boast that its oil was 'effectual in all affections of the mucous membrane of the stomach and bowels, inducing a feeling of repose and tranquillity'. Uncle Vern had obviously found this to be true.

One morning I woke up feeling so sick that I felt I had to stay in bed. Uncle Vern eventually came looking for me and when I told him I was sick he went off and returned with a small bottle of his blue mallee oil. 'Take this,' he said. 'It'll fix you.' I asked him what I should do with it. He said, 'Swallow it, rub it on or bung it up your arse. Whichever you do, it'll fix you.' After Uncle Vern left the room, I thought I might swallow some of it, but then I realised I was feeling a little better, so I ended up not using the oil after all. I did keep the bottle, though, thinking it might come in handy one day. About six months later my cousin Jill Baker, Vern's daughter, had a twenty-first birthday party in the woolshed. Most of us drank too much at the party, and next morning one friend in particular was extremely sick with a hangover. I remembered I had the blue mallee oil and suggested he take some. He felt sick enough to try anything, so when I poured him a small quantity of oil in a glass, not much more than thimble-full, he downed it in one gulp. His eyes bulged. 'I think it's burned my guts out,' he gasped. It did seem to relieve his hangover, though.

At the time I lived and worked with him, Uncle Vern was in his early fifties. He was physically strong and a very hard worker, so he took some keeping up with. Most of the older men I worked with in those days, like Vern, had been through the Depression, so they were accustomed to doing it tough. Yet, despite his rough bush ways, Vern had come from a fairly privileged background. He had gone to one of Melbourne's more exclusive schools, Wesley College, and had rowed for Wesley in the Head of the River, the big inter-school regatta. Uncle Vern died of a heart attack at 66. He was feeding sheep one Saturday and suddenly dropped dead. This was in 1971. Everyone thought Vern had probably been working too hard. He was very kind to me at a time in my life when I needed some support, and I remember him with affection.

Vern's property had a beautiful location on Corio Bay near Avalon, the site today of one of Melbourne's airports. Even then it should have been obvious to everyone that 1800 acres at such a location was potentially very valuable. Some people thought he made a bad mistake when he bought the place, because it was swarming with rabbits, but he gradually got on top of the rabbit problem, and he made some extra money by supplying shell grit for the trotting track at the Melbourne Showground. Seven years after Uncle Vern died the place was sold for a lot of money to the chemicals company ICI.

I think I was at Uncle Vern's place no more than three or four months, although it seemed much longer. It was during that time, thanks in part to Vern's influence, that I finally decided that farming was the life for me—that I would make the land my career. It was an easy decision to make in one sense, for my family's sheep property, Lanark, was there for the taking. But I was still not quite ready for Lanark. For one thing, I had to do National Service, the several months of military training that every young Australian male in the 1950s was obliged to do. I enjoyed National Service. I liked the camaraderie, the sense of belonging to a group who were all in the same boat. Boys at boarding school have the same kinds of fellowship.

If nothing else, National Service represented a high point in my athletics career, for I was the 19th Battalion's 100 yards champion. But it was not the absolute high point. Just before I went into National Service, I worked for a while in the wool stores at Geelong, a job that Uncle Vern had arranged for me. I was there to learn about wool classing and I worked in the reclass section, which meant my main job was carrying wool from one bin to another. I did learn a lot about the business, though, thanks largely to a delightful old character named Mr Baird, the manager of the reclass section. I boarded with a widow who was a good cook and lived in a pleasant house that was close enough to the wool stores for me to walk to work.

Wanting something to do after work and at weekends, I began training with the Geelong Guild Athletic Club, which happens to be the club that the great middle-distance runner John Landy belonged to. I was still well above average at sprinting—I think my best time for 100 yards was 10.5 seconds wind-assisted, which was not bad for a club athlete—but, no matter how hard I tried, I was never able to make it to the top in the sport. I used to think I might have made it if I had been blessed with longer legs, but that was probably wishful thinking.

While I was at Geelong, Melbourne's Oakleigh Athletic Club invited the famous Australian sprinter, Hector Hogan, to compete in a 100 yards handicap race. It was a twilight meeting, and the field was lit by a few lights, which was fairly unusual at that time. Athletes from the Geelong Guild were invited to the meeting too, and I found myself racing against Hogan. Hogan started from scratch, of course, and I think I started from about ten yards, but I remember him flashing past me before we got to the halfway mark. Regardless of whether Hogan won the race, which I am pretty sure he did, I have ever since been able to boast that I raced against one of the world's finest sprinters. It is worth boasting about, too, because Hogan at one time shared the world record for both the 100 yards and 100 metres. He was one of the favourites for a gold medal in the 100 metres sprint at the 1956 Olympics but had to settle for a bronze. I was shocked a few years later when I learned he had died of leukaemia at the age of 29.

7

TREES ACROSS THE TASMAN

Working on Uncle Vern's farm had convinced me once and for all that my future lay in farming. Before going home to take over Lanark, however, I felt I needed to widen my farming experience. I wanted to see if there were other, better ways of running a sheep property, and the best place to do that, I decided, was New Zealand. I cannot now remember exactly why New Zealand farming methods interested me. Perhaps it was simply because I knew that they did things a little differently there and that they also did them successfully.

I consulted my uncle, Bob Strachan, about it. Uncle Bob was married to my mother's older sister, Norma, and was a prominent businessman in Hamilton. He thought the idea of my going to New Zealand was a good one, and he gave me a letter of introduction to the pastoral company Goldsborough Mort, which had offices in New Zealand. So I flew to Christchurch and walked into the Goldsborough Mort office there on a Friday afternoon. This was early November 1957. The man I spoke to read Uncle Bob's letter, looked me over and said he could fix me up with a job on the best farm in Canterbury by Monday.

He was as good as his word. By Monday morning I had started working there. The farm, called Erskine, was near Kirwee, a town west of Christchurch. It was owned by Ian Young, whose family had come from Erskine in Scotland. Ian's father had been a potato farmer near Invercargill, but after his father died Ian decided to sell up and buy land in Canterbury. The country he settled on was fairly light and tussocky, and it would probably have been considered rather ordinary by Canterbury's high standards. Yet by applying his expert farming knowledge Ian was able, over time, to work it up into a wonderfully productive property. It was so productive, in fact, that although he had no more than 1500 acres (600 hectares) of land he managed to turn off around 8000 prime fat lambs a year, plus large quantities of lucerne hay and small-grass and clover seed.

The first thing I learned about New Zealand is that its farmers were miles ahead of us in just about every area of farming. For all I know, this may still be true. I found New Zealand farmers to be models of efficiency. It seemed to me they had just about perfected every facet of rural production. Their attention to detail was one of the things that impressed me. The last job I did for Ian Young was to work a paddock over and over, sitting in an old Lanz Bulldog tractor. By the end (I know this because I kept count) I had worked the paddock fourteen times. Ian Young was producing high-value Italian rye grass seed in this paddock. Working a paddock fourteen times seemed to me then a waste of time and money, but I realise now that Ian knew what he was doing. The paddock was like a bowling green, and he made a lot of money from the seed he produced there.

Why were New Zealand's farmers better than ours? One reason, I think, was that New Zealand's agricultural college system was in some ways more advanced than ours. The main reason, though, was simply that the country's superior soil and more reliable rainfall allowed farmers to operate in much smaller areas, which in turn meant they could afford to devote far more time and attention to each acre they farmed.

I spent only six months at the Youngs' property, although I learned so much there and did so much it felt like a couple of years. Ian had a Dutch immigrant working on the place, who lived with his Dutch wife in a separate cottage, and I slept there, too. In the evenings, though, I always had dinner with the Youngs and was made to feel part of the family. I became very fond of them. If nothing else, I owe them a debt of gratitude for the knowledge I acquired in my six months there. The things I learned on the Youngs' farm had a lasting influence on my own approach to farming.

One of those things was how farmers could improve their properties by planting trees. Ian had begun to plant rows of trees as windbreaks and shelter belts. As I recall, he was planting Mexican cypresses (*Cupressus lusitanica*). Typically, he prepared the ground well and planted the trees with a high degree of professional skill. Most of them survived and grew successfully. We planted a couple of long rows of Mexican cypresses while I was there. The rows were two or three trees wide plus, I think, some shrubby plants on the windward side.

The trees were planted not only for shade and shelter. Already, a few farmers in New Zealand were starting to realise that the trees they planted today could eventually be a valuable source of timber. Ian Young was certainly thinking along those lines, which is why he planted the trees in configurations that would enhance their value as timber. It was a new idea to me, but I found it interesting. I filed it away in my memory for future reference.

8

CICELY

I had another reason for going to New Zealand. In a way, my trip there was a silent protest. Let me explain.

Cicely and I had been aware of each other in a vague sort of way since childhood—our families knew each other and sometimes met—although we were certainly not childhood sweethearts. She was just a girl that I would see every so often at functions and get-togethers but did not take much notice of. After all, she was three years younger than me. Her father, Patrick Gaussen, had a large grazing property named Vasey Farm near Balmoral, west of the Grampians. The property had been owned by the Gaussen family since 1925. I formed my first romantic association with a Gaussen girl when I was a teenage agricultural student at Longerenong College, although it was not with Cicely but her older sister, Jill, who was about my age. To call it a 'romantic association' is really to overstate its importance, for the truth is it was just a teenage attachment of a harmless type that was common then.

That was 1953, when I was eighteen. My next contact with a Gaussen girl was in 1956, and this time it was Cicely. It happened this way. A good friend of mine from Hamilton, Graeme Gunn, who was an exceptionally talented footballer, had been recruited by the Melbourne Football Club and had moved to Melbourne to play there. I sometimes went to Melbourne to spend the weekend with him, camping in the half-house that Graeme and a few others rented in South Yarra. One evening I tagged along with Graeme and his girlfriend when they decided to go to a kind of nightclub in the city. We were walking along Bourke Street, just in front of the Myers store, when we came upon Cicely, who had probably just been shopping and was with a group of girlfriends. Graeme apparently felt sorry for me because I did not have a female companion, and on seeing Cicely he grabbed me and said in my ear, 'Look—there's that Gaussen girl, Cicely. Ask her to join us.' I was reluctant, guessing she would probably turn me down, but Graeme insisted, so I went up to Cicely and asked her to come along. As I had feared, she declined.

She must have given me her phone number, though, because not long after this I asked her out again and this time she said yes. Soon, I was taking her out regularly, and before long I was smitten. This was 1957 and Cicely was now nineteen years old. Then came some worrying news. Cicely was an accomplished equestrian, and her parents had decided to send her to a riding school in Ireland called the Burton Hall Riding Centre, which was run by a memorably named ex-soldier, Colonel Dudgeon. The news was worrying for me, because I half-suspected that the real reason her parents wanted her to go to the riding school there was that they hoped she would meet and marry a gentleman-farmer. But there was nothing I could do to stop her going, for everything was arranged. Cicely had been booked into the school and was due to sail to England in October 1957.

It was clear to me that I needed to take pre-emptive action to ensure I did not lose her. An opportunity arose in September 1957, when we were in Melbourne together for the weekend. Cicely stayed there with her grandmother, Christine Thornley, and I again stayed with Graeme Gunn and his mates. Cicely and I had already talked about getting married in a roundabout sort of way, but this weekend I made sure the talk became action. I took her to Hardy Brothers, an old firm of jewellers in Melbourne, and bought her an engagement ring. I chose the Hardy Brothers shop because that is where my father had bought my mother's engagement ring. In fact, my father may well have bought the ring from the same elderly jeweller who sold the ring to us. He remembered my father as 'Mr Dave'.

Then came the awkward part: breaking the news to Cicely's parents. It had already been arranged that Cicely and I would drive to their property that evening, and as we travelled up in my mother's Volkswagen I felt myself becoming more jittery by the mile. We found Mrs Gaussen cooking a roast in the kitchen. 'You're very late,' she said. I explained it had been a big day—and I blurted out that Cicely and I wanted to become engaged. Mrs Gaussen said, 'But you haven't asked Mr Gaussen,' the implication being I should do so immediately. So I walked rather hesitantly down the hall to what was called the smoking room. He was in there, reading a newspaper and sitting in front of a fire blazing with red gum logs. I told him about our intention to become engaged. He did not seem unduly surprised, but he replied firmly that, no, he would not agree to it. So I returned to the kitchen where Cicely was waiting. Her mother said, 'What did he say?' I told her he had said no. 'What bloody rot!' she said to me. 'I knew your father well, and he was a fine man.' Then she strode up the hall to the smoking room. A few minutes later, she emerged and declared, 'You will announce the engagement tomorrow.'

So we became engaged. A month later, Cicely left for England. I did not want her to go, of course, and I resented the fact she was going. After all, we were merely engaged, not married, and I was still afraid that she might meet somebody else there and call our engagement off. It was against this background that I decided to go to New Zealand. I suppose I was making a point: I was showing her that she was not the only one who could flit off overseas. So I set off for New Zealand in November 1957, just a few weeks after Cicely's departure.

I returned to Australia in June 1958. Two months later Cicely came back, too. We married a little over four months later—on 8 January 1959—and moved at once into our home, the house at Lanark that had been built in the mid-1940s to accommodate the property's manager. By now, I had been living at Lanark off and on for several years in a dwelling known as The Hut. I was certainly there in late 1956 when the Olympic Games were held in Melbourne, because I remember listening at Lanark to Olympic events on the radio. Built half of weatherboard and half of fibro, The Hut had one bedroom, a small kitchen with an old wood-burning stove, and a long-drop dunny outside. The Gunn brothers, Noel and Murray, had built it for me behind the manager's house in 1956 when I began living at Lanark intermittently. The Hut has been put to various uses since then, and for the past twenty years or more it has served as an office.

During the years I had been staying, off and on, at Lanark, I had managed to plant a few trees around the house and a few individual trees around the property. The manager of the property, Keith Fort, helped me plant them. The trees were planted simply to provide shelter for the house. They were mainly Monterey cypresses (*Cupressus macrocarpa*), plus a few natives. The latter were mainly tuarts (*Eucalyptus gomphocephala*)—a coastal eucalypt native to the southwest of Western Australia. We grew the tuarts by direct seeding, which at the time was an unorthodox way to grow them but has since become fairly common practice. The method I used was quite straightforward. I shallow-ploughed the strip of ground where I wanted the trees to grow, spread the tuart seed and raked it in. It was necessarily a simple process for it replicated, more or less, the process by which trees are seeded in nature. Many graziers in the Western District used this same method in the 1930s to plant long rows of sugar gums (*Eucalyptus cladocalyx*).

In retrospect, I can see that even then I was becoming keen on the idea of planting and nurturing trees. An incident that occurred not long after I planted the tuart seed illustrates this. Keith Fort had bought a number of turkeys, which he intended to fatten and eat. I called in one day when he was away and saw that one of his turkey gobblers and a number of turkey hens had got into a fenced-off area where I had sown the tuart seed. The gobbler was scratching at the seed as hard as he could and spreading it everywhere, which made me see red. I grabbed a stick and threw it at him with the intention of driving him out of the tuart row. Unfortunately for the gobbler, the stick spun like a boomerang and struck him in the neck. He ran for a short distance then fell down dead. I buried the turkey and said nothing to Keith Fort. I reasoned that an admission of guilt would not bring the turkey back and was sure to strain my relations with Lanark's live-in manager. Perhaps Keith concluded the turkey had been taken by one of the wedge-tailed eagles that nested in drooping she-oaks (*Allocasuarina verticillata*) nearby.

Right from the outset, I kept a daily journal of things that happened on the farm. Here is an entry from around that period: 'I've had my eye on a new site for a house since about Christmas time. Mum had a look at it the other day and also G C Gunn. Both thought it was okay. Makes use of a view of the swamp and the Lanark Lake. Cicely thinks it's great too.' What I find interesting about this journal entry is that I used the word 'lake'. This was 30 years before Lake Cicely really came into being, and I was already dreaming of creating a large body of permanent water in the vicinity of the house. Another entry records the fact that as far back as April 1957 I bought a rotary hoe for use in planting trees and shrubs—another indication that I was already taking the idea of planting trees seriously.

In May 1958 Keith Fort left Lanark to take up a stockman's job at Yarram Park, a well-known property near Dunkeld, which made it possible for Cicely and me to move into the manager's house at Lanark after we were married eight months later. The house was made of weatherboards and had an iron roof. It had two small bedrooms, a sitting room and a tiny bathroom that was heated by a chip heater with an iron chimney. The toilet was a long-drop dunny outside. Our water supply consisted of two 1000-gallon galvanised-iron tanks that were fed from the roof. We did have running water in the house, but there was not enough water pressure to run a shower. The house had no electricity as such, either. The only power we had was supplied by a 32-volt battery unit that I charged periodically using an old engine in the shearing shed. It was one of the excellent Ronaldson Tippett engines that used to be manufactured in Ballarat.

There was still one employee at Lanark—a local man named Jack Childs. He had started working there while Fort was still manager, and he stayed on when I took over. Jack lived in Branxholme but used to camp in The Hut during the week.

By the time Cicely and I took up residence in the former manager's residence at Lanark, the house was surrounded by young trees that were growing vigorously. Tree-wise, it was not much of a start—but it was a start.

9

A FADED SQUATTOCRACY

I must say something about the character of the part of Victoria where Lanark is located. I do so not for the benefit of Victorian readers, all of whom would be well aware of the Western District and the kind of place that it is (or, perhaps, that they imagine it to be). People who live elsewhere in Australia, though, tend to have only a vague idea of where and what the Western District is. They may know that Malcolm Fraser's former property, Nareen, is there, but that is quite possibly all they do know about it.

Allow me then to provide a personal view of this region, surely one of the most interesting in the whole of rural Australia. The Western District is most simply defined as the southwestern corner of Victoria, extending roughly from Ballarat in the east to the South Australian border in the west. It was once considered the heartland of Australia's so-called squattocracy, and there was a good reason for that. The district occupies a geological feature known as the Basalt Plain, which is a more or less flat expanse of rich volcanic soil that extends from Melbourne to the far side of Hamilton. In fact, the Basalt Plain starts to peter out at Lanark; we have the tail end of it. Not far to our west the basalt soil is replaced by lighter soils with limestone outcrops.

Thanks to its excellent soil and suitable climate, the Western District is arguably the best sheep-grazing land in Australia, and it became the nation's premier wool-growing region. Even today, the area is said to produce around 15 per cent of Australia's wool clip. The sheep farmers who took up large tracts of land there in the 1800s became wealthy—and thus was born the legend of the district's squattocracy, a legend that the media has kept alive to the present day. Just the other day a friend drew my attention to a description of the district's sheep graziers that he had found on some website or other. We were called 'those pampered, moleskin-wearing, Jaguar-driving, privately educated Western District squattocracy types'. The writer intended to be humorous, of course, yet the image he presents is still widely accepted as fact.

People can make jokes about the Western District and its squattocracy, but the truth is there is not much to laugh about. Those old-time graziers with big properties may have been wealthy, but their wealth enabled them to support and hold together entire rural communities. Take Branxholme near Hamilton, which I consider my 'home village'. Until the 1930s, virtually everyone who lived in Branxholme had a job. Branxholme residents were either employed by the owners of Bassett, or they were paid to provide services in the district. There was virtually nobody in Branxholme of working age who was not gainfully employed in one way or another. Today, I would guess that half the people who live in Branxholme cannot find permanent work of any kind—and that is not through any fault of theirs.

Thus, far from being funny, the story is actually a very sad one. So what went wrong? Well, the district suffered two body blows in the first half of the last century. The first was the loss of so many of its finest young men in World War I. As far as I know, no historian has so far managed to assess the true impact on Australian society of the loss of such men, especially in country towns. In the Western District, as elsewhere, the war wiped out a significant portion of a whole generation of movers and shakers—men who otherwise would have gone on to lead the way in the rural industry. After all, the men who went to that war were all volunteers, and I think it is fair to assume that, by and large, they were the men with the most energy and initiative.

Take the case of Hamilton. In 1914, when the war began, the town's population would have been no more than 5000. (In the last census before the war, in 1911, its population count was 4902.) Yet from those 5000 or so residents, 374 men and 32 women enlisted for military service, and 79 of these men were killed. I would guess that those 374 men represented a very high proportion of Hamilton males of eligible age, which meant males in their late teens and twenties. This, in turn, shows how devastating for a town of Hamilton's size was that loss of 79 lives. Clarke Street in Hamilton has the questionable distinction of being, reputedly, Australia's most fertile source of World War I volunteers. There were 33 houses in Clarke Street then, and 25 young men living in the street opted to go off to war.

The other body blow for the Western District—and the whole of rural Australia, for that matter—was the Depression in the 1930s, when many big grazing properties went to the wall. Even for the properties that survived, things were never the same again. Farming methods changed. Fewer and fewer farm workers were employed. Farming became more intensive and depended heavily on pasture improvement. This was, I believe, the end of the squattocracy as everyone knew it. Why? Because the men who might have had the drive and ingenuity to adapt successfully to the new circumstances that the Depression created in the bush were lying dead in war cemeteries in Europe.

In the event, the grazing industry was unable to adapt and, as a consequence, went into decline. The reality is that for many years now the district's graziers, like most Australian farmers, have struggled to survive financially.

After the first *Landline* program about Lanark appeared in 1991 I suspected that some viewers, knowing that we were sheep graziers in Victoria's Western District, may have assumed that the environmental work we did there was just a personal indulgence by members of the old squatter class. I hope I make it absolutely clear elsewhere in this book that nothing could be further from the truth. The environmental work we did at Lanark was certainly not an indulgence. It strained us to the limit financially.

However much the district may have changed, Hamilton remains a very livable, handsome, comfortable country town. For the past 30 years or so Hamilton has had a fairly static population of around 10 000, which is not to say the city is in any way stagnant. One effect of having a stable population like this is that the people who live in Hamilton tend to know everyone else, so it is a well-integrated community. Unfortunately, the population outside the town has not been so stable, and various changes in land usage, most notably the spread of blue gum plantations, have had a lot to do with that. In certain parts of the district, especially around Branxholme, the plantations have replaced farms, and the farmers, their families and employees have gone—a tragic development in my opinion, which I will discuss later in this book.

For all that, Hamilton and the district around it remains a marvellous place to spend one's life; I know of no place better. Over the past 50 years I have been to Melbourne hundreds of times, yet each time I return I still get a lift when the Grampians come into view and I know I am nearing home.

Hamilton has been called the wool capital of the world, although wool is not its only claim to historical fame. Hamilton also has a place in Australia's aviation history. In 1936, young Reg Ansett (he had just turned 27) started out in the airline business by running a service between Hamilton and Melbourne with a six-seater Fokker plane. Five years earlier he had begun using a second-hand Studebaker car to ferry people from Maryborough to Ballarat and from Hamilton to Ballarat. A few Hamilton people invested in Ansett's airline venture, including Billy Beggs, a brother-in-law of my father.

Beggs was an interesting character. Born in 1879, he had varied business and rural interests, he bred racehorses with fair success, he was elected a Hamilton councillor, and he was president of both the Hamilton Club and Hamilton Golf Club. Beggs was reputedly one of three Hamilton people who put up money to get Ansett's first plane in the air. All three later became directors of Ansett's company. Beggs remained a director until 1956, and he was the company's chairman for several years. Reg Ansett once declared that, but for Beggs, Ansett Airways might not have existed. He was referring to his role in dealing with a crisis that arose in 1937 over the delivery of aircraft. The odd thing is that, by all accounts, Beggs never once flew in a plane.

10

LANARK AS IT WAS

I was barely 22 years old when I took control of Lanark, and only 23 when I took up full-time residence there with my new bride. This was probably just as well. If I had been much older, I may well have been disillusioned by the sorry state the place was in and lost heart. As it was, I was young enough to relish the challenge of turning Lanark into a first-class property. This proved to be an enormous job, although I did not realise it at the time. That was probably just as well, too.

The country around Lanark looked very different, of course, when white farmers first came there with their sheep and cattle. It was then an open woodland with a dense growth of trees and associated under-storey plants along the natural drainage lines. These trees would have included drooping she-oaks, manna gums (*Eucalyptus viminalis*), blackwoods (*Acacia melanoxylon*), swamp gums (*Eucalyptus ovata*), woolly tea-tree (*Leptospermum lanigerum*), silver banksias (*Banksia marginata*), which were called 'honeysuckles' by the early settlers, and black wattles (*Acacia mearnsii*), which are the pioneer wattles of the district. The farmers got rid of most of the trees. Black wattles were targeted in particular because their bark is rich in tannin, which made them useful in the tanning process.

Most of the land clearing in our area took place in the second half of the 1800s, especially the 1870s. We have a first-hand description in 1886 of how rapidly the old Bassett sheep station, of which Lanark was part, had been stripped of trees and other vegetation in the previous fifteen years. It appeared in an article in the *Hamilton Spectator*, the local newspaper (which celebrated 150 years of publication in 2009). The writer had this to say:

> I was not prepared for the transformation the place has undergone. The forests have almost disappeared and with the exception of the bleached weather boats in trunks of the larger gum trees, there is nothing to indicate they ever existed; the main branches have all fallen and cleared away, whilst the thick honeysuckles [banksias] have vanished, root and branch. Vast lengths of drains have been completed to link up with the creek and the land is all now healthy.

Let me describe what Lanark was like when I took it over in 1956. To begin with, it was almost tree-less. Apart from some pine trees and two stands of sugar gums that had been planted in the 1930s as shade for the stock, Lanark was a bare expanse—a moonscape almost. The pastures were neglected and degraded. There was a patchy covering of native grasses but little else. The stock were in poor condition, too. The sheep had foot rot when I arrived there, and the cattle were old and looked like bags of bones.

Years ago an RMIT publication quoted me saying this about Lanark when I took it over:

> Looking back, I often wonder why the hell I didn't sell the property. All the water had been drained. It was absolutely bare, apart from the thirty-five pine trees planted along the railway line during the Depression, 300 pines that my father planted in 1939 and two very small sugar gum plantations on the property. Every other tree I have the opportunity or privilege to walk under I have planted myself or an extremely close person to the family has planted.

Clearly, the absence of trees at Lanark was a problem I could not fix quickly. Trees take a long time to grow—frustrating when you want them to grow quickly. But I was confident it would not take me long to fix most of Lanark's other problems. I believed I could turn things around fairly quickly. In fact, my first management decision was not a good one. With hindsight, the most sensible method of dealing with the foot rot would have been to sell all the sheep and start again. At the time, though, I was convinced I could beat it—which I did manage to do eventually, although it took an immense amount of time and effort. I had to separate the sheep with foot rot from the rest of the mob, and pare their feet (by no means a pleasant job, given that the feet were often fly-blown), which I then soaked in a bath of formalin. Moreover, this all had to be repeated periodically.

In 1956, Lanark had no remnant vegetation. By this I mean that none of the original vegetation, including trees, that grew there before farmers started clearing the land had survived—apart from a few degraded areas of native grassland that had been grazed continually for 100 years or more. The only remnant vegetation visible on the place was over the fence—along the railway line that forms Lanark's southern boundary. There were some silver banksias, manna gums and black wattles growing beside the line, plus some patches of native grass. Not long afterwards, however, the railway authorities decided to poison the lot because they felt the railway line would then need less maintenance and be a better fire break. (In my opinion, they were wrong on both counts.) It took them several years of spraying to get rid of this indigenous vegetation, but eventually all of it was gone. Today, exotic plant species grow along the line, including phalaris grass, self-sown pines and numerous weeds. These have to be sprayed annually to maintain access and ensure an effective fire break.

On our side of the fence, meanwhile, I was busy establishing a row of pine trees about 150 metres from the boundary. The pines were meant to provide shelter for my sheep—but they were also meant to mask the sight and sound of the steam trains that used to run fairly regularly along the line. Today, if steam trains still ran along that line, I could probably charge people to come and watch them.

These pines were my first major planting. I am happy to say they are still there today.

11

TREES AND MORE TREES

I am sometimes asked how and when the idea of planting trees on a large scale at Lanark first entered my head. People assume there must have been a moment of conversion when I saw the light—when I was suddenly transformed into a farmer with an environmental conscience. But there was no such moment. The idea of large-scale tree planting came to me bit by bit in response to a variety of influences. For instance, Cicely believes that when I started planting trees at Lanark I was really reacting to the bleakness of the landscape. After all, I had not grown up on the property. If I had, the look of the place would probably not have bothered me. Instead, I spent my formative years living in a comfortable family home in Hamilton, surrounded by a large, leafy garden, so when I did go to live at Lanark I found the bleakness of the place hard to accept.

Cicely may well be right about this, but looking back now I can see that even as a young farmer I did have a concern for the environment, even if I was not conscious of it at the time.

In 1959, for instance, I attended a meeting called to establish a Hamilton branch of the Field Naturalists Club of Victoria. I became a founding member of the branch, served on one of the committees and periodically went on field trips with other members, so I did have a particular interest in the natural world. Certainly, very few other farmers of my age belonged to organisations devoted to the welfare of the environment.

Around the middle of 1956 my good friend Graeme Gunn, who as well as playing football in Melbourne was studying to become an architect, came to spend a weekend with me at Lanark. The main house on the property was still occupied then by Lanark's manager, Keith Fort, so Graeme and I camped in The Hut, which at that time had neither electricity nor running water. It would have been a fairly spartan weekend at the best of times, but it happened that this weekend was also unbearably hot. There were no trees in the vicinity to speak of, so The Hut was the only place we could find shade. But it was so hot inside under the iron roof that we might just as well have stayed out in the sun. We were also short of water. Even if The Hut had been equipped with running water, there would not have been enough water for us to have a shower. Graeme said to me, 'What a bastard of a place! You've got to do something, Fenton. You've got to get some water. The bloody stuff falls out of the sky. Why don't you catch some of it? And why don't you plant some trees and cool the place off a bit?'

Graeme's complaint registered with me. It was not long after this that I began planting the shade-and-shelter trees around the house that I mentioned earlier.

Another incident sticks in my mind. Soon after Cicely returned from England in August 1958, her father drove down in a ute from his property near Balmoral, which is between Hamilton and Horsham, about 80 kilometres away. It was to

Top left: My uncle Jack Fenton (far right) at a family dinner in Hamilton held to farewell him before he went off to war in 1915. Jack was killed in 1918, just months before the war ended. Although Jack looks cheerful enough about his impending departure, the rest of the family are clearly despondent. Standing, from left: my father, Dave Fenton, Aunty Nell (Dave's sister), Marcus Whiting, Bill Beggs, John Fenton and Aunty Mollie (Davy), Uncle Jack. Seated, from left: Helen Fenton (nee Laidlaw), Aunty Jean (Whiting), Ken Beggs and Ken's mother, Aunty Gwen. *Courtesy of Bruce Maloney*

Top right: Jack Fenton (right foreground) in a parade down Collins Street, Melbourne, before his unit's departure for France.

A photo dating to the early 1920s of my father, Dave Fenton, in his cricket whites and an old Geelong College blazer. *Courtesy of Bruce Maloney*

My father (middle) with his chauffeur, Herbert Bailey (right), and a friend, Bill Brake, during a sheep-branding operation at one of the grazing properties owned by my grandfather John Fenton. This photo must have been taken in the early 1920s, because Bill Brake died from the effects of wartime gassing within four years of his return from France. He was with Jack Fenton at the gas attack. *Courtesy of Bruce Maloney*

A brief romantic attachment: the England cricketer Herbert Sutcliffe with Aunty Nell in the back garden of Lantana, the Fentons' family home in Hamilton. The boy in the photo is Nell's nephew Ken Beggs. *Courtesy of Bruce Maloney*

My parents, David and Lesley, with me astride a horse at Tahara South (one of old John Fenton's properties), 1937. *Courtesy of Bruce Maloney*

With my parents at our home in Chaucer Street, Hamilton, 1940. *Courtesy of Bruce Maloney*

My sister, Robin, and I in the surf at Bathing Box Beach, Portland, in January 1944. *Courtesy of Bruce Maloney*

Lanark in the 1950s presented a desolate scene.

My first visit to the nursery at Wail, 1951.

One of the ill-fated willows I planted near a swamp at Lanark in 1957.

Cicely at the time of our engagement, 1957.

Life certainly wasn't meant to be easy: Cicely hanging out the washing not long after we came to live at Lanark.

In our first few years at Lanark we ran not only sheep but cattle—like the Herefords I am feeding here from the back of a ute in 1959.

Jack Childs, the station hand at Lanark, lamb marking in the early 1960s. As with the photos above, Lanark's landscape was still largely treeless.

My new bride driving cattle at Lanark, 1959. Just beyond the rise visible in the photo is the site of the future Lake Cicely.

Cicely at Uncle Vern's property at Corio Bay, 1961. Geelong is visible in the distance.

Cicely and children, 1963. Amanda was then three years old, David and Patrick two and Johnnie eight months.

Top left: Paddy Gaussen with his grandsons David and Johnnie Fenton at Vasey Farm, 1965.

Top right: My youngest son Johnnie and I (with Christmas the dog) take a core sample from the site of Lake Cicely's empoundment bank to be tested to ascertain clay quality and depth. *Courtesy of Jim Sinatra and Phin Murphy*

Above: The three Fenton boys—Johnnie, Patrick and David—with a few of the black wattles that reappeared where I deep-ploughed near the Other Side Swamp, 1965.

Left: My three sons catch the last fish in the fast-disappearing swimming hole at the Other Side Reserve during the 1967–68 drought.

The six images on the following pages trace the history of the wildlife dam that I built and fenced off in our front paddock in 1969. It proved to be the very last dam I built at Lanark.

1 Contractors 'cut the core' of the new dam—that is, strip off the topsoil to expose the clay beneath—in 1969.

2 I supervise the construction of an island in the dam—an operation that necessarily requires some precision.

3 Water flowing through a contour drain starts to fill the dam in 1970.

4 My partners in Rural Trees Australia, Roger Young and Murray Gunn, planting indigenous trees as habitat near the dam wall, 1984. A tractor driven by Roger pulls the tree-planting machine with Murray inside it.

5 Black swans on the dam, 1996. Three-year-old oak plantings are visible beyond the dam.

6 A second dam above the wildlife dam and linked to it by indigenous habitat.

be his first look at Lanark, which was soon to become his daughter's home. Cicely came with him. I was waiting at Lanark to greet them, and when the ute pulled up I stepped forward, opened the door on the driver's side, where Mr Gaussen was sitting, and said, 'Welcome to Lanark, sir.' In those days I often called him 'sir'. He stepped out, looked around slowly with a frown and said to me, 'John, if you're intending to bring my daughter to this godforsaken place to live, you'd better plant some trees.'

When it came to planting trees, Paddy Gaussen knew what he was talking about. Unlike most other Australian farmers at that time, he believed in planting and nurturing trees for their own sake. He had done this at his own property, Vasey Farm. He had made a special effort to foster the growth of river red gums (*Eucalyptus camaldulensis*), the finest of all eucalypts in that part of Victoria, by fencing off areas where they could survive and flourish. As a result, Vasey Farm today has some of the best stands of red gums in the district.

By now, my mind had been opened to the potential benefit of planting farm trees, even if I still did not have the motivation to start doing it. In September 1958, not long after this visit, something I heard on the radio gave me the motivation I needed. It was a talk on the subject of farm trees by someone from the NRCL (Natural Resources Conservation League of Victoria), which had grown out of the old Save the Forests Campaign to promote the revegetation of Victoria. I knew next to nothing about the NRCL then, but I was soon to learn that it was a splendid organisation staffed by dedicated people, one of whom was the man I heard on the radio, Lance Edgar, the manager of the NRCL's nursery in Melbourne.

I was so impressed by the radio talk that day that I decided to go to Melbourne to speak to him personally. The NRCL was based at Springvale, a suburb of Melbourne, so I booked myself into a new motel at Oakleigh, which happened to be one of the first motels, if not the very first, to open in Victoria. I invited Cicely to come with me and stay at the motel, but as we were not married she would have none of it. She did come with me to Melbourne, but she stayed with her grandmother.

The next morning I went to the NRCL's office. I arrived before opening time, feeling quite excited about being there, so much so that I took a photo of the front entrance, which I still have. I met Mr Edgar in the nursery. I bombarded him with questions about trees—which to plant and when, where and how to plant them. I came away believing that, where trees were concerned, Lance Edgar was the ultimate expert. He seemed to know everything about them. I felt he took a special interest in me, too. At that time, few farmers were planting trees, and I imagine he was pleased that a young farmer like me (I was 23 at the time) had driven all the way from Hamilton to speak to him about it.

I returned to Lanark not only with a lot of information about trees but with a resolve to start planting them. I also had about 50 native trees to plant, which I had bought at the league's nursery. They were tube stock—that is, plants in tubes—although the tubes then were made not of plastic but of plywood, which is why they were known as 'veneer tubes'. The 50 trees were meant to be a trial planting. The nurseryman had recommended 25 different types of wattles and gums that he thought might grow well at Lanark and supplied me with two trees of each species, the idea being that I would plant them all and compare how they grew.

As far as I can remember, only three of those 25 tree species were indigenous to my own part of Victoria. At that time, selecting trees indigenous to the area where you wanted to grow them was an idea that had not taken hold. Even then, though, I did have an interest in the concept. Not long after this I visited the National Herbarium at the Royal Botanic Gardens in Melbourne and asked for a list of trees that grew naturally around Branxholme. (As I later discovered, the list I was given was not entirely accurate. There were two or three trees on the list—sallow wattle [*Acacia longifolia*] was one of them, I remember—that ought not to have been there.)

It was not until 20 years after this, in the late 1970s, that I really became wedded to the idea of planting indigenous trees by preference. An obvious reason for doing so is that an area's indigenous trees will naturally provide a better habitat for wildlife indigenous to the same area. The trees themselves do better, too, partly because they are compatible with local soil fungi. It has been found that most Australian trees have a symbiotic relationship with local soil fungi. The trees supply the fungi with carbon, and the fungi supply the trees' roots with nutrients and water, thus promoting the trees' growth and ensuring their health. On the other hand, a tree planted in an area where there is no compatible fungi is likely to struggle. Just the other day I read of eucalypts planted in China that did badly until the cause of the problem was identified: the trees did not have the right soil fungi.

As suggested, I planted my 50 trees in two rows not far from the house, making a careful note of which species were planted where. The idea was to see which trees did well and which did not. In fact, about two-thirds of them did badly, from which I concluded they were the wrong species for our locality. In fact, as I now know, there may have been other reasons for their failure to thrive. For one thing, I did not do enough to control weeds. Nowadays, I start preparing the ground two years before planting. The key thing is to eliminate competition: you need to make sure there are no live plants near the trees you are planting. Today, we spray the planting area beforehand with a herbicide at least twice to make sure the trees that we are planting have it all to themselves. As I have said elsewhere, successful tree planting depends on three essentials: species selection, preparation and protection.

I also made a mistake by planting those 50 trees in autumn. In those days, my corner of Victoria used to have wet winters quite regularly, which meant that any small trees you planted in autumn would probably have to sit in damp clay for five months or more. They would also have to survive numerous frosts. As a result, many trees did not survive. Some years later we abandoned the idea of autumn planting and instead began planting in spring. Late September to early October is the preferred planting time for present conditions at Lanark.

About a third of the 50 trees did grow pretty well, however. Manna gum (*Eucalyptus viminalis*) was one of them, and drooping she-oak was another. So I now had some idea which trees to plant if I ever got around to planting them. I also made a start on trying to improve the property environmentally. The first project I undertook was the rehabilitation of a swamp at the back of the house, known as the Racecourse Swamp because, in the days when the area was part of Bassett Estate, racehorses were trained there. I fenced off an area beside the swamp and planted around 60 willow trees (*Salix* spp) at intervals of 20 metres. It was my first attempt at a multiple planting of trees away from the house and it proved a dismal failure: the swamp was not permanent. It filled every three years or so but dried out completely between times. It was certainly too dry for the willows, which died one by one until none was left.

Lest anyone reading this should visit Lanark one day and go looking for the Racecourse Swamp, I should explain that its name was changed later to the Other Side Swamp. This is how the change of name occurred. At that time there was a small dam in front of the house which, for safety reasons, was strictly out of bounds for our children. They were not allowed to go near the dam, much less swim in it. Instead, we swam in a pool in the Racecourse Swamp, which was on the other side of the house dam. We would pack the children in the car—we never walked there, even though it was only 400 metres or so away—and tell them we were going to the 'other side' to swim. Soon, 'other side' became the name of our destination, and not long after that Other Side was adopted as the name of the swamp, too.

My visit to the NRCL nursery in Melbourne proved to be a watershed event for me, for it triggered a keen, active interest in planting trees. The interest continued to develop. In April of the following year, 1959, I drove to Wail, a hamlet in the Wimmera district, about nine kilometres from Dimboola, to buy more plants from a nursery there. At the time I went there, the nursery was growing a lot of trees for roadside planting in the Wimmera, one of the main cropping areas of Victoria, which had been largely cleared of trees many years earlier. While the absence of trees may have made cropping easier, it created a depressingly bare landscape. The Wimmera is extremely flat, so you could look in every direction and hardly see a single tree breaking the horizon. This was bad for the land and, I suspect, bad for the farmers who lived there. A pleasant landscape may not earn farmers any extra money, but it does improve the quality of their lives.

According to my farm journal, I bought a 'mixed box of native plants', plus 125 pine seedlings from the Wail nursery that day. I also had a long yarn about trees with a forester there, Trevor Arthur. He spoke to me about the many benefits of planting trees on farms and offered practical advice on how to go about it. His basic message was that trees provided stock with shelter and shade, both of which were essential for their welfare. As best as I can recall, he said nothing that day about trees being good for the environment, although that would probably have been taken for granted. After all, he was speaking not to someone from the city but to a young farmer who had to make a living from the land, so he would naturally have pointed out the financial advantages of growing trees rather than the environmental ones.

What he had to say went down well with me. To some extent he was preaching to the converted—or to the half-converted, anyway. I was now convinced, if I had not been convinced before, that planting trees on a large scale at Lanark would be good both for the farm and for the environment. I returned from my visit to the nursery at Wail full of big ideas about where and how to plant the thousands of trees that I had already made up my mind to plant. For the time being, though, the big ideas had to wait. There was just too much else going on—too many other things to worry about. Making a living from the farm was one of them.

12

NO SPARE TIME FOR FOOTY

From the day Cicely and I arrived at Lanark as a newly married couple, I had worked flat out to get the sheep property running efficiently and, as I hoped, profitably. The fact that children had begun to arrive made this all the more important. In fact, the children came in a rush. The first of them, Amanda, was born in December 1959, barely eleven months after we were married. Then, in January 1961, came the twins, David and Patrick, and finally, in August 1962, our third son, John. Cicely and I then had four children under the age of three.

Life was so busy that I felt obliged to give up playing footy for Branxholme, my local team. I loved playing football, and I continued to miss it long after I played my last game. A cousin of mine, John Whiting of Bassett Station, gave up playing football at the age of 30. He told me later that he always regretted doing it. I quit at 24, so I missed it all the more. I gave up football not merely because I had so much else on my plate; I was also conscious of the fact that I was responsible for a young and growing family. I was now running the sheep farm on my own. A bad injury on the football field would have landed me in deep trouble financially if it meant I was laid up for months.

Once I had given up playing football I found myself with some spare time—and I devoted nearly all of it to planting trees. At that time I was buying most of my trees from the Wimmera Forest Nursery. It may have been a 90-minute drive away, but it was an extremely well run place and I liked the people there. In particular, I became friendly with the district forester, Bill Middleton. Bill was the son of a doctor at Dimboola, who sent him to board at Geelong Grammar, no doubt believing he would not have a future at Dimboola. In the event, Bill became a forester and ended up back in his home territory.

Bill Middleton was a quiet-natured man, yet he held very strong views about trees and their place in the environment, views that had a big influence on my own ideas on tree planting. As young as he was, he had seen a lot of bushland cleared, and I know he took this to heart. It was a cause of immense sadness to him. As the district forester from 1959 to 1976, Bill did some outstanding work reafforesting roadsides in the Wimmera. Some of this work was experimental: he advised the Country Roads Board of Victoria on which species of tree to grow and then monitored how each of them fared. In this way, he was able to determine the right trees for the right location. He also encouraged farmers in the Wimmera to plant trees around their homes, which quite a few did.

Bill and I shared an interest in birdwatching. He knew a lot about birds, and he was happy to share that knowledge with me. He and I, plus one or two other friends such as Murray Gunn, used to go camping together, observing and recording bird life. Some of these camping trips were in the superb Little Desert nature reserve in the heart of the Wimmera, southwest

A familiar combination: me with trees to plant. *Courtesy of Jim Sinatra and Phin Murphy*

Dr John French's 'coppice-with-standards' plantation he persuaded me to establish in 1985.

Graeme Gunn (left) with Cicely and our son Johnnie, shortly before Johnnie left for England in 1985.

An example of pines growing in conjunction with a permanently fenced indigenous habitat component, 1991. As the photo shows, red gum sleepers were still used then in railway tracks.

The aerial photos on the following pages, taken in 1992, show the result of 35 years of environmental rehabilitation at Lanark. Instead of a barren, windswept expanse, there were now 100 000 or more trees and extensive wetlands. *Photos supplied by David Hay and Don Thomson and taken by Lindsay Stepanow, Ballarat*

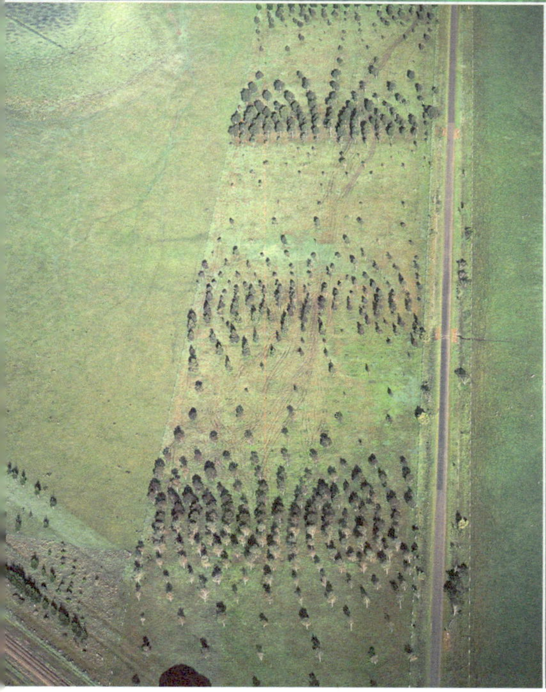

Bill Funk arrives at Lanark in 1993 with a ute full of oak seedlings ready for planting. In the background are, from left, Cath and David Fenton, Michael Shenks and Rochelle Ruddock. Michael and Rochelle later married.

My daughter Amanda with one of the trees she helped me plant in 'her plantation' at Lanark in 1962.

Below: Shorn sheep shelter from the wind in 1992 behind the plantation of mainly she-oaks that Amanda and I planted 30 years earlier.

Rod Bird (centre) with two South American agricultural scientists who had come to inspect our environmental work at Lanark.

I was delighted when Jim Sinatra (centre), and Top End grazier Bob Purvis (right) from Atartinga Station visited Lanark.

These RMIT students seem interested enough in what I have to say about a cypress plantation at Lanark. *Courtesy of Jim Sinatra and Phin Murphy*

The Hut. Left to right: my brother-in-law David Peddie, David Fenton and children, Murray Gunn and a visiting English ornithologist.

of Dimboola. The name 'Little Desert' is misleading; although much of the area is arid, it abounds in native life, both animals and plants. It is not little, either, extending from Dimboola to the South Australian border. Yet in the 1960s the area was in serious danger of being cleared for farmland. This crazy plan had the support of influential politicians, too, most notably Sir William McDonald, the Minister for Lands, who in 1967 called a meeting to discuss it. At the meeting, held at Kaniva, just north of the Little Desert, Bill Middleton and others spoke strongly against the idea. Indeed, the plan to open up the Little Desert caused general outrage. Melbourne's newspapers ran countless letters from incensed readers. Faced with so much opposition, the Victorian Government scrapped the plan a few years later and instead declared most of the Little Desert a national park.

Graeme Gunn, by now a qualified architect, was another who gave me advice (sometimes whether I wanted it or not) on where to plant trees at Lanark. As far back as the early 1960s Graeme used to stay with us every year or two, usually for a weekend, and on each of his visits he and I did a lot of reminiscing about old times over numerous bottles of beer. He always made a point of inspecting the trees I had planted since his last visit, and he was often critical of what I had done. 'Why in God's name did you plant them there? They will look bloody awful. You'll have to pull them out.' And so on. As best as I can recall, I never pulled out trees that I had planted, but I did take his advice on certain design matters. There are features of Lanark's layout today that can be traced back to advice that Graeme gave me 50 years or more ago.

It would be wrong to imagine I devoted several days a week to planting trees. After all, I had a large sheep property to run and a family to support. In those early years at Lanark,

I probably worked around 48 hours a week on the farm and maybe spent five hours a week planting trees. In effect, I spent all my spare time planting trees. It was as much time as I could afford.

The truth is I knew very little then about trees. I even had trouble spelling their names. Recently, I came upon an entry that I made in my farm journal in March 1957, in which I recorded the fact that Cicely had suggested 'banksia' as the name for a new paddock I was developing. In my journal I spelt the name 'bankshia'.

In that same month, March 1957, I reported in my journal that I had been enrolled as a paid-up associate member of the NRCL of Victoria, which entitled me to a 15 per cent discount on trees bought at the league's nursery at Springvale. Another entry twelve days later shows that I took advantage of the discount by buying 200 trees. Later that year, according to my journal, I was busy erecting fences to protect the rows of trees I intended to grow. Planting trees had clearly become an important part of my working life.

I planted trees in my spare time right through the 1960s. I even planted them in 1967, one of the worst drought years on record in Victoria. The drought began in autumn that year and continued, with just an odd break, until early the following year. Jack Childs was still helping me run the place, but after the drought really set in I could no longer afford to pay him; Jack had to leave. It was an amicable parting, and Jack had no trouble finding another job.

The 1967 drought did not hurt us as much as it did other farmers, however, one reason being that I offloaded a large number of sheep just in time. In those early years I occasionally engaged a farm management consultant to advise me. He visited Lanark in September 1967, some months after the drought had begun, and told me I needed to reduce stock numbers immediately. I had a mob of 1600 wethers which he said I should shear at once and sell as quickly as I could. I disagreed, arguing that in two months' time the wethers would have a full growth of wool, which would make them worth a lot more. The consultant insisted: if I waited two months it would be too late. So, rather grudgingly, I took his advice and sold the wethers at a reasonable price. It was one of the best decisions I was ever talked into making. Two months later, by which time the drought had really set in, I could hardly have given the wethers away.

By now, I had begun planting in spring, usually October, as a matter of policy, and spring in 1967 was horribly dry. Yet even the trees I planted then did pretty well. In fact, most of the trees I planted in those early years survived, which was just as well. If a sizeable number had died on me, I am sure I would have given up large-scale tree planting as a lost cause. This is why anyone setting out to plant trees should make sure at the outset that they are going about it the right way, thereby ensuring the success of the planting. If you do not have success with tree planting early on, you will find it hard to keep going.

13

ROOM FOR WILDLIFE, TOO

When they got rid of the native vegetation at Lanark, they also got rid of the native wildlife. So in 1961, I fenced off some land, including a swamp, with the idea of creating wildlife habitat by getting the original vegetation to grow there again. I was told that seed from the original vegetation might still be in the soil, waiting to spring into life if only the right conditions could be created. According to one theory I heard, the trick was to burn the land, which would get old seed in the soil to germinate. According to another theory, the thing to do was to plough the land, which would bring old, long-buried seed to the surface, where it would germinate. I decided to do both. I burned half the fenced-off land and deep-ploughed the other. Where I burned, nothing grew. Where I ploughed, a few black wattles appeared.

The land in question had been cleared as far back as the 1880s and had been bare-grazed for 80 years at least, but it was possible that before the land was cleared a few black wattle seeds had fallen into the cracks that tend to open in the ground when it is hot and dry and that this seed, having been brought to the surface by the plough, had now germinated. Whether or not this was true, I had convinced myself that, in a practical sense, there was no chance of Lanark's original vegetation regenerating. If I were to get trees and other vegetation growing there again, I would have to plant them—at least in the first instance. I knew that, later on, there would be regeneration from the trees I had planted, which is precisely what did happen.

So I went ahead and planted. I began planting native species that I believed to be indigenous to the area. In other words, I was trying to make good the loss of remnant vegetation. But the information I had obtained on what used to grow at Lanark before European settlement proved to be only partly accurate. As a result, I introduced quite a few species to the area that were not indigenous to it. These have since regenerated of their own accord, so in a sense my initial mistake has been perpetuated. But was it really a mistake? That is debatable. True, I had introduced 'foreign' species to the area. On the other hand, I did add to the land's biodiversity.

This is what makes it so difficult to come up with an accurate figure for the number of trees Cicely and I have planted at Lanark over the past half century. True, I can look at the numerous areas that I fenced off and planted, and then do a reasonably accurate estimate of the number of trees growing there. The trouble is that many of those trees are the product of regeneration, having grown from seed produced by trees that I planted. Regeneration of this kind often results in dense thickets around wetlands, with a virtually uncountable number of stems.

Trees were not the only plants we had to reintroduce at Lanark. The smaller, under-storey plants had long disappeared, too, and they had to be brought back. One of these was sweet bursaria (*Bursaria spinosa*), a native shrub that is usually thorny. It is an extremely valuable plant in the Australian environment because its flowers are a key source of food for certain wasps, one of which preys on the scarab beetles that do a lot of damage to eucalypts. In other words, this wasp protects the eucalypts by keeping the beetles under control. Sweet bursaria is also a haven for small birds. It was therefore a plant that I was particularly keen to reintroduce at Lanark.

The scientists who established the link between the health of eucalypts and the presence of sweet bursaria did a great service to the environment. They include Dr Rob Davidson, who led a CSIRO team of researchers at Armidale, in New South Wales, and Dr David Curtis of Greening Australia, also at Armidale. The two worked closely together. Their findings emerged from a study of eucalyptus dieback in the New England district of New South Wales.

Ideally, a habitat reconstructed on a farm should consist of the entire suite of plants that once lived there, grown from the nearest available seed source to the farm. When I went to reintroduce sweet bursaria, however, I found there was no seed source nearby. The species had been eliminated from the local district, apparently because early farmers considered it a weed. I did find a seed source eventually, but it was 30 kilometres away.

My advice to anyone intending to set aside an area for native planting, whether indigenous or not, is to forget entirely about its appearance. You should certainly not set out to create something that looks good. On the contrary, you should expect to end up with something that looks untidy, for untidiness is in the natural order of things where native Australian habitats are concerned. The Australian artist John Olsen made the same observation in an interview with *Good Weekend*'s Janet Hawley.

'The Australian landscape is not a gentle place,' he said. 'It lacks the formal sense of geometry you find in Europe. It arranges itself like undisciplined graffiti, like a blue cattle dog's hind leg.'

Yet Australians as a whole strive mindlessly for tidiness. Perhaps it is something we have inherited from our British forebears. Personally, I have a special dislike of the Tidy Towns competition. It would be much better to have an Untidy Towns competition, for in my experience untidy towns tend to be happy towns. Similarly, untidy farms tend to be sustainable farms. Some years ago there was a move to create a Tidy Farms competition. To my immense relief, it did not get off the ground. Cicely once bought a framed sign which later hung for many years in our kitchen. It read: 'This house is tidy enough to be healthy—and dirty enough to be happy.' The same could be said of the environment.

The Australian landscape as a whole, in its native state, is naturally an untidy one. So forget about pruning trees or cutting back vegetation from fences. You should also forget about picking up stones, for stones are part of native habitat, too. If you pick up the last ten stones lying in a paddock, there is a good chance you will have got rid of the last skink lizards there, because skinks use the stones as cover.

Dead and dying trees are an essential component of any natural habitat. This is why a complete ecosystem cannot be created overnight. A large proportion of Australian birds and animals, most notably parrots, owls and bats, nest in tree hollows or under peeling bark. As we know, tree hollows do not develop usually until the trees start to senesce—that is, start to show signs of ageing—and this may not happen until the trees are over 50 years old. So habitat creation is necessarily a long-term process. Indeed, Neville Bonney, an ecologist friend, has estimated that the first trees I planted at Lanark may not start to senesce until they are 130 years old.

For the same reason, farmers should never get rid of old trees dying in their paddocks just because they are old and dying, for dying trees are sure to be home to a variety of wildlife. In any case, it is in the farmer's interest to leave the trees there. The bats who live in them are known to play a key role in controlling pasture and crop pests, but the bats operate within a limited radius of their home base—maybe only 100 metres or so. So if the farmer gets rid of an old hollowed-out tree standing in a paddock, the paddock may no longer be protected by bats. Not long ago I heard of a farmer who had been growing winter crops in a paddock next to a stand of old trees inhabited by bats. He found the crops in that paddock to be almost entirely pest-free.

If you simply cannot bear the look of dead tree trunks and branches spread about a paddock, perhaps because the paddock is close to the house, then push them into a heap and maybe plant a tree in the middle of them. The dead wood in the heap will still be available for animals and insects as a habitat, and the heap will protect the tree you have planted until it is big enough to stand on its own. I must add a couple of provisos here, however: you should not heap fallen timber in the way I have just suggested unless rabbits in your area are under control and unless you have satisfied yourself about fire hazards.

Sometimes a bit of tinkering can vastly improve the look of a wildlife reserve. The wetland that we converted into Lake Cicely—a haven for water birds—was only about 100 metres from our house, although my architect friend Graeme Gunn would have liked it even closer. After we created Lake Cicely in 1967, Graeme used to say to me, only partly in jest, that I should burn down the house and build another one close enough to the lake to overlook the water. As the next best thing, Graeme urged me to plant trees on the far side of the lake. He said this would create an optical effect, making the lake look as if it were closer to the house than it really was.

I opted to plant manna gums, a noble indigenous species, at the top of the slope there. Manna gums are relatively fast growers and reach a considerable height when mature. On the shoreline of the lake, in front of the manna gums, we planted woolly tea-tree, another indigenous species, plus other riparian (water-side) shrub species. As the trees grew, they began to produce precisely the optical effect that Graeme had said they would: they seemed to pull the lake towards the house. The taller they became, the closer to the house the lake looked to be. Ever since, we have called it the Graeme Gunn Plantation.

14

NO NEED FOR CHEMICALS

As well as planting trees, we embarked on a pasture-improvement program at Lanark soon after I took over. As a result, the property's carrying capacity rose steadily. In the 1950s we used to shear around 4000 sheep on average each year. In 1966, we managed to shear around 10 000, which was almost six to the acre. (I should point out that we did not shear as many as this regularly. We hit the 10 000 peak in one year only.) Being young and eager to learn I was always open to advice on how to make the property more productive. In the early 1960s I accepted one bit of 'expert' advice that I wish now I had ignored. I was told I could fix Lanark's insect problem by spraying the place with DDT.

Dichlorodiphenyltrichloroethane or DDT has long been regarded as one of the most harmful of farm chemicals. Not only was it judged to be carcinogenic, but it was found to accumulate over time in greater and greater concentrations in the environment. Accordingly, it has been banned in Australia since

1987. In the 1950s, though, it was considered an indispensable weapon in the battle against insect pests. Many heavily stocked graziers, like me, used DDT to kill grubs that killed grass. One of my neighbours dosed the whole of his property with DDT by spraying it from a plane. I doubt if a single insect, good or bad, was left alive on the place. Suburban gardeners used DDT, too. I do know that various rose societies used to urge rose growers to use it. I have a booklet, *Introduction to Rose Growing*, published in 1954 by the National Rose Society of Victoria, which recommends the use of DDT to kill aphids.

In 1963 I read Rachel Carson's recently published *Silent Spring*. Carson was an American biologist and author who was years ahead of contemporaries in her understanding of how various pollutants could have serious harmful effects throughout the wider environment. Indeed, she has been credited by some with starting the environmental movement that was to sweep through the Western world in the 1970s. In *Silent Spring*, Carson focused on the damaging effects of DDT. Her essential message was that DDT did not merely harm the insects it was intended to kill; she provided evidence to show that it harmed the entire environment, bird life especially. This explained the title of the book: without birds, spring was silent.

Carson's book made a big impression on me. As it happened, I had been feeling unwell for several weeks, and when I read the book I began to wonder if the diazinon that I used in my new sheep spray race might be the cause. I certainly did not use a mask, glasses or other protective gear while handling chemicals like this. Few farmers at that time did. I was also under fairly severe pressure of work, trying to cope with the huge number of sheep we were running. About that time I attended a short how-to course for young farmers organised by the Marcus Oldham Agricultural College at Geelong. During one class the lecturer got onto the subject of stocking rates. He asked, 'Who here is carrying three sheep to the acre?'—and a lot of hands went up. 'Who's carrying four to the acre?' A few hands went up. 'Is anyone carrying five to the acre?' Only one or two hands went up, so the lecturer said, 'I'm sure nobody is carrying more than that.' I then put up my hand and said I was carrying six. 'I'm sure you are,' the lecturer said with a smile. He obviously did not believe me. But it was true: we were carrying six sheep to the acre, and it was a considerable strain for all concerned.

I tell this story to demonstrate again that, contrary to what a lot of people think, I did not come from a 'green' background. As a young farmer I sprayed parts of the property with pesticides and other chemicals, including DDT, and I was quite prepared to stock the property with six sheep or more to the acre in the hope of making more money, even though this was obviously to the detriment of the land—and the detriment of my family, too. At the same time I was doing all this, though, I was also planting shelter belts of trees. So on one hand I was over-taxing the land, and on the other I was doing my best to restore it. The truth is I was still feeling my way. I had yet to work out a consistent approach to managing the land.

At that time, of course, the science of farming was relatively primitive. Farmers were still using remedies from the dark ages to treat sick animals, as an entry in my farm journal in 1956 shows. It records the fact that a twelve-month-old steer of ours was suffering from impaction, a serious condition that develops when cattle eat certain types of feed, especially onion grass, that form a ball inside their gut. I had treated the steer twice with an off-the-shelf drench, but this had not worked. Lanark's manager, Keith Fort, then administered what I described in my journal as his 'cure-or-kill' drench. It consisted of mineral turps, kerosene, soap and fat. This did not work either, so I consulted the manager of a rural supplies store in Hamilton. He suggested another recipe: mixing two pounds of Epsom salts with two ounces of ground ginger, then adding two parts of boiling water and one cup of treacle. We tried his concoction, again without success. As my journal notes without comment, the steer died two weeks later.

I stopped using pesticides at Lanark in the second half of the 1960s—either 1967 or 1968, as best as I can recall. Pesticides became unnecessary when bird numbers multiplied. Birds are insect predators, after all, and there were enough of them now to keep insect numbers in check. I stopped using superphosphate at Lanark about the same time. It had become clear to me that I needed to pull back from the mad quest for greater and greater production that I had been engaged in. It was not doing me, the farm or the environment any good.

True, superphosphate and other chemical fertilizers made the grass grow, but it appeared to me that things at Lanark were generally in a downwards spiral. Our sheep-farming operation was becoming more intensive in every respect: more drench, more fertilizers, more costs, and more and more work. Cicely was now doing a lot of farm work. She had been ever since Jack Childs left. It amazes me still that she managed to do as much work as she did while remaining an attentive and nurturing mother for our four small children.

I sensed that I needed to take the pressure off everything and everyone. On the one hand, this meant reducing stock numbers. On the other, it meant cutting right back on fertilizers and chemicals. Other farmers who heard I had given up using artificial fertilizers like superphosphate viewed me as a kind of heretic. It was accepted as gospel then (as, generally speaking, it still is today) that to farm productively without using super, potash and similar products was impossible. All the experts—government agronomists and the like—took this for granted. I found it impossible to obtain advice about the pros and cons of supering, because nobody accepted there were any cons. I used to speak every so often to a local Department of Agriculture agronomist. When I asked him, as I sometimes did, about the feasibility of farming without superphosphate, he had a standard reply: 'John, just spread a few bags of super around and you may manage to earn enough money to plant more trees.'

In fact, Lanark's level of production held up after I stopped supering, thanks to the residual effect of the super previously spread there, but then slowly declined. Against that, I was satisfied the environment at Lanark was healthier and, of course, I no longer bore the financial burden of having to buy the super. Since my son David took over the running of the property he has begun supering again, although in a strategic, controlled way. David believes he has got the balance right—the balance between the need to maximise production and the need to keep the environment healthy. In other words, he believes he has developed a sustainable agricultural system. I agree with him.

This is not to deny, of course, that there are plenty of organic farmers and others who are running successful farming operations without super. Most Australian farmers, though, are farming their land today as they have always farmed it. Only one thing has really changed: farmers are working their land today even harder than farmers did, say, half a century ago when I got into the business. Farmers feel they have no alternative if they are to have any hope of covering costs. In the end, though, it is a self-defeating cycle. Because you are struggling to make ends meet, you start to farm more intensively—more stock, more fertilizers, more chemicals—to try to earn more.

And so the cycle goes on. All the while, the base resources of your land are being steadily exhausted.

As I write this, the Wimmera district, one of Victoria's main cropping areas, has again been experiencing dust storms. Farmers are seeing countless tonnes of their soil blown away. In my view, dust storms in an area like the Wimmera are a direct consequence of farmers working the land too hard. It is hard to blame them for that, though. In Australia's fickle climate, and against a background of ever-rising production costs, theirs is a tough, ongoing battle: to obtain a financial return that is at least big enough to allow them to stay on their land and in their homes.

15

A WETLAND RESTORED

In early 1966 Cicely and I began by far the biggest project that we had undertaken so far at Lanark: the construction of a large dam to provide a permanent water supply. In effect, we were setting out to install a wetland to augment the already reinstated Racecourse Swamp. In the process, we would also create a habitat for birds, especially water birds, and we wanted that habitat to be as varied as possible in terms of both plant types and water levels. The more varied the habitat, the more sustainable the habitat would be. The project was to provide practical benefits, too—most notably a permanent water supply for the house and garden as well as for the stock. If the need arose, we would have water to protect ourselves against fire. Lanark's number-one problem when Cicely and I came to live there was a shortage of water, both at the house, where we had only a 1000-gallon tank, and in the paddocks. This problem, we hoped, would be fixed once and for all.

At the time we were preparing plans for Lake Cicely, we were fortunate to have the services of Michael Cook, our farm consultant, known to us as Mick. He was in favour of the various environmental projects we were undertaking, which was pretty unusual for a farm consultant at that time. Mick suggested we go one step further and use the water in Lake Cicely for irrigation, which required some modification of our original design for the lake. Incorporating irrigation in the project did provide a bonus: it convinced my bank manager to lend us the money we needed to see the whole project through.

What the dam ended up doing, in fact, was to create an expanse of permanent water that we later named Lake Cicely. The lake forms a kind of horseshoe around the house and, when full, measures about seventeen acres (seven hectares) in area and holds about 60 acre-feet (74 000 cubic metres) of water. The site for the lake was just below the old swamp—that is, further down the catchment. The construction work went ahead so steadily that the entire project was finished by the following summer—that is, the summer of 1966–67. For that, we can thank the skill and efficiency of the contractor, the late Kevin Shepherd, who had as many as five Fordson tractors working on the job at one time. Kevin also had two vans stationed on the site. One served as a workshop and the other as a mess hut. In all respects it was a big operation.

With the dam wall now in place, we all looked forward to seeing water banking up behind it in the winter of 1967. That proved to be a false hope, for 1967 turned out to be a terrible drought year. Indeed, the 1967–68 drought was one of the worst in Victoria's history. According to the Bureau of Meteorology, the first six months of the year were the driest January-to-June period on record in the twentieth century across large areas of Victoria, and as far as I know that included our own district.

Until the record was beaten in 2007, Melbourne had its driest 365-day period in 1967–68. The rainfall in that period totalled 318.0 mm—or only twelve and a half inches. In 2006–07, the record-lowering total was 316.4 mm.

As a consequence, hardly any water appeared behind our big new dam in the whole of 1967. True, the dam wall did develop some small leaks, as dry dams tend to do, and these later needed fixing. But it was the drought that kept the swamp dry initially. Water did not start to flow into it until eighteen months after the dam's completion—after the drought broke in 1968. Before long, Lake Cicely began to appear, as if from nowhere, and the lake has never been totally dry since.

Because the lake was intended to be a habitat for water birds and other water-loving flora and fauna including frogs, swamp rats and white-tailed water rats, we fenced off the whole area from stock. We knew this would also allow a full suite of indigenous riparian plant species to establish themselves there. Maybe it was appropriate that the fenced-off wetland at Lanark should have been installed by someone named Fenton. I have read that the family name is derived from the ancient English word 'fen', which means 'marsh', and from the equally ancient English word 'tun', which means a fenced-off enclosure. Together, they describe perfectly the project we undertook to reinstate Lanark's swamp. The fenced-off reserve around Lake Cicely continued to expand in the decades that followed. By the time my son David finished fencing off the latest, and probably last, addition to it three years ago, the reserve had grown to 65 acres (27 hectares).

The old swamp above Lake Cicely had been drained many years before by previous owners of the property. There would have seemed nothing wrong with that at the time: it was what farmers all over Australia did if they had swamps or wetlands on their land. Why did they drain them? One reason was to acquire more grazing land. The main reason, though, was to reduce or, better still, eliminate the risk of certain diseases that are prevalent in wet areas. Liver fluke, a common disease in cattle and sheep, is one of these. Snails are an intermediate host of the parasite that causes the disease, and snails flourish in and around swamps.

The soldier-settlement schemes that operated after both world wars led to much more intensive farming, which, in turn, encouraged people to drain more and more swamps. In Victoria, and probably in other states too, local drainage boards were set up to promote and oversee the drainage of wet areas. There was a real sense of compulsion about it. Draining wet country became as imperative as spraying weeds. Everywhere you looked in the early 1950s you saw machinery hard at work, digging drains.

I believe the damage done to the environment by this mania for draining natural wetlands and swamps was catastrophic. It was not just the swamps and wetlands that were destroyed, either. One farmer in the Branxholme area drained sections of his property and in the process destroyed the only pristine creek in the district. He diverted the creek through a channel, which effectively got rid of all the plant species that had lived along the creek. The creek had been a habitat for platypus, a whole suite of bird species and at least one endangered species of fish, the pygmy perch. There were tiger snakes there too. By the time he finished his draining operation they were all gone—including the tiger snakes, which promptly migrated to my place and set up permanent home in the wetland I was then re-establishing. Previously, when the property was bare and dry, tiger snakes were never seen at Lanark. About the only snakes we used to see were the less aggressive copperheads.

As far as I can tell from the records, the wetlands at Lanark were first drained in the 1880s, when Lanark was part of Bassett Estate. Farmers throughout Victoria—and across much of Australia, for that matter—were draining swamps as long ago as that. Over time, these drains gradually became blocked and the swamps reappeared, which is no doubt why the drainage boards were set up after World War II. Even today, you can see the outline of the drains quite clearly in aerial photos.

In 1956 I decided to reinstate the old Racecourse Swamp. At that time, drains were still channelling water out of the swamp, so these had to be filled in. We wanted to restore the old swamp to its natural state, the condition it was in before white settlement—with the same supply levels. Now, swamps in Australia tend to dry out periodically; some swamps dry out every year. The Racecourse Swamp dried out every few years. This is a perfectly natural occurrence. Australian wetlands have a naturally alternating sequence of wetting and drying, and it is wrong environmentally to try to prevent a swamp drying out by increasing its capacity. A far better option is to construct a new permanent body of water on a suitable site.

This is what we set out to do when we built Lake Cicely below the Racecourse Swamp. Our objective was to create a sizeable body of *permanent* water—a small lake, really—that water birds could use as a habitat at any time of year and around which riparian plants could become permanently established. The site was ideally suited to such a construction. It consisted of a large, fairly flat basin with a gentle gully falling away to the northwest. All we needed to do was build a wall to dam the water. We removed the topsoil along the line of the wall we intended to build, exposing the hard clay underneath. We then raised the wall by piling more and more clay on top of the exposed clay, compacting it all as we went along. When we finally had a wall of clay of the required height—about four metres—we covered it all with the topsoil that we had removed at the outset.

The wall, when complete, extended in a slight arc for a distance of about 100 metres. We finished building the dam in about six months. Initially, to keep our bank manager happy, we called it the Stockyard Irrigation Dam. Only later did we rename it.

Cicely and I bore the entire cost of the project. What I found incomprehensible at the time—and I still find incomprehensible today—is that other farmers in the district who drained their properties received government money to do it. They were paid subsidies to install the drains and, in my opinion, damage the area's environment. Yet I could never get a cent from any government authority to support the work I was doing to improve the environment at Lanark by planting trees and reinstating a habitat for water birds. All the early tree planting we did at Lanark was in our own time and at our own expense.

Initially, our dam project was viewed with suspicion by certain authorities. The concern was that we might be blocking a natural water course. Some people from the old Victorian Soil Conservation Authority (a fine organisation, incidentally) came to see for themselves. They agreed it was not a natural watercourse, so we were allowed to proceed with the dam's construction.

As it happens, we do have a permanent creek running through the property, but it is a long way from Lake Cicely. My guess is that until the mid-1800s—that is, before the country at Lanark was first cleared for farming—there were numerous springs in the vicinity. The springs may well have been there for thousands of years, but they would not have survived for long after the arrival of cattle and sheep. I do not understand the mechanics of the process, but I know from observation that springs that get trampled on repeatedly by sheep or cattle tend to dry up.

Initially, after creating Lake Cicely, I did use some of its water for irrigation. The irrigation system we set up was the so-called Keyline system developed by an Australian engineer, Percy Yeomans (better known by his initials P.A.), a great innovator in farmland management. Under this system, water is channelled through a drain that has been dug along an elevated contour. When the drain is blocked—usually with canvas sheets that are known in the business as irrigation flags—water overflows across the land below it. I irrigated in this way for about three years but then stopped. The reality was that the catchment feeding the lake was relatively small, so the lake did not fill up with water every year. Effectively, this meant we would not be able to irrigate in dry years, which was precisely when we needed to irrigate, so the whole exercise ceased to be workable.

Jim Murphy, an irrigation specialist, referred to this in a paper he wrote about irrigation by Western District sheep graziers in the 1960s. Murphy was employed by the Victorian Department of Agriculture to advise farmers in the district on irrigation, a position he held from the early 1960s to the mid-1980s. This is what he had to say about Lanark:

> The Fentons toyed with the idea of irrigation even to the extent of constructing Lake Cicely, but did not proceed further. Was it an opportunity lost? Arguably yes, but there were sound enough reasons for their decision. It was a first-class irrigation site and 40 acres [sixteen hectares] could have been irrigated from the 60-acre-feet [74 000 cubic metres] storage.

> However, John and Cicely were committed to raising a young family and to wresting a living, without hired help, from their 2000-acre [800-hectare] property. Irrigation would have required more intensive stocking, additional management commitments and a greater labour input.

If Lake Cicely is not fed by a creek or springs, where does it get its water from? The answer is simply from run-off after rain in the local catchment above it. Today, the run-off is markedly smaller than it used to be, one reason being there are now many more trees growing higher up in the catchment than there used to be—and trees reduce run-off. Not only are there several shelter belts of trees that we have planted on our own land, including a substantial agroforestry windbreak of conifers and eucalypts three kilometres long, but, beyond that, there is now a large blue gum plantation on a neighbouring property. Whatever else may be said about them, the fact is that blue gums do trap and consume a lot of water.

Years ago we found an interesting clue to the likely presence of springs at Lanark in the distant past. On a dry hillside we discovered a number of Aboriginal axe heads. Clearly, groups of Aborigines had spent a lot of time at this spot over the years. The question was why. Only 200 metres away, at the bottom of the hillside, there was a creek that ran for most of the year. Why did the Aborigines not camp on the bank of the creek where water was available? The conclusion I came to was that there had once been a spring where the axe heads were found, which supplied the visiting Aborigines with all the water they needed. The reason they gathered at the spring, rather than on the bank of the creek below, was that the spring was a vantage point. From there, they could see if there was any game in the area. Perhaps they could also see approaching enemies. The spring presumably disappeared when farmers first cleared the land and introduced hard-footed stock.

As I write this, Lake Cicely has been attracting and fostering bird life for well over 40 years, which to me seems quite a long time. How remarkable then to think that local Aborigines used the very same water resources, on the same site, for almost a thousand times as long as that. The evidence suggests that Aborigines have inhabited the southwest corner of Victoria for 35 000 years at least, and maybe for 40 000 years. These local Aborigines, various clans of the Gunditjmara nation, tended to base themselves in one place, travelling only when it suited them. In summer they walked on age-old tracks down to the Portland coast to eat mussels and crabs on the beach. For most of the time, though, they lived on and around Lake Condah, a wetland teeming with wildlife.

They built stone houses there, the foundations of which still exist. They also built and managed extensive networks of ponds at different levels, in which they farmed eels. The eels were caught in an intricate system of traps and smoked in ovens fashioned from burnt-out trees, after which surplus eels were traded with other tribes over a vast area that possibly extended beyond the Murray River. All this happened tens of thousands of years ago. Against a time span of that magnitude, everything we have done at Lanark seems very recent indeed.

There is a sad contrast between the way Aborigines managed the land sustainably over a period of 40 000 years or more and the way it has been managed by the European settlers who displaced them. In barely 200 years (170 years in the case of my own district), Europeans have succeeded in turning a healthy landscape into one that is clearly ailing. I recognised this as a young farmer, and ever since I have sought to learn about and understand the land management techniques that the Aborigines employed. As one of my senior Aboriginal friends, John Lovett, says, 'We don't need your sympathy, but we do want your understanding.'

16

THE RETURN OF THE BIRDS

In some ways, I take more pride in the return of birds to Lanark than I do in anything else that has happened on the property in the past 50-odd years. I use the word 'return' deliberately: the birds had been there before. By all accounts, the country around Branxholme, which includes Lanark, used to teem with bird life before white settlers arrived. In 1858, a remarkable character named James Bonwick (1817–1906)—teacher, school proprietor, gold prospector, historian, archivist and journalist—had this to say about our area: 'The Smokey River rises from the Branxholme Swamps. Great numbers of native companions, herons, ducks and geese frequent these swamps. The country between Branxholme and Hamilton, eighteen miles, I found to be like that from the two falls—plateau forests and swamps.'

It has been estimated that when I took over Lanark in 1956, 80 per cent or more of the 200 bird species that had been present in the area before white settlers arrived were no longer there. By clearing the woodlands and draining the swamps, farmers had deprived them of a habitat. Today, the bulk of those 200 species have returned. This is a great result for the environment as well as for the birds, for an abundance of birds is always an indicator of a healthy landscape. I had this pointed out to me by an officer of the World Wide Fund for Nature in 1985 not long after the fund appointed me a trustee. I made a note of what he said: 'The number of species of birds present on a given piece of farmland and the number of individuals present on that land is a direct indicator of the health of that piece of land.'

The first list of bird species at Lanark was compiled more than 60 years ago by Murray Gunn, a naturalist with a passionate interest in birds, and an older brother of my friend Graeme. Murray was only a teenager then, but in the late 1940s he happened to spend some time at Lanark working for his father, a builder, who was erecting a woolshed there. While Murray worked at Lanark he made a list of the birds he saw in a notebook that he always carried with him. Years later, Murray and I were both members of the local branch of the Victorian Field Naturalists Club, and I have a distinct memory of talking to him one day about the importance of having enough bushland on the farm where birds could breed. It was probably the first time I discussed the matter with anyone.

The most detailed of all bird life studies at Lanark was undertaken in 1987 by Elizabeth Jacka, an RMIT student. Drawing from records compiled by Murray Gunn and from our own farm diaries, she was able to compare the number of bird species believed to have been in the district in 1840, before white settlement, with the number observed in 1956, when I took over Lanark, and the number in 1976, after the wetlands had been reinstated. Her conclusion was that the 200 species estimated to have been present in 1840 had fallen to 39 by 1956 but had risen to 133 by 1976. This was a remarkable increase, especially considering that by 1976 the revegetation of Lanark was far from complete. Although we had done some woodland planting by then, Lanark's overall woodland component was still quite small, amounting to not much more than a fortieth of the property. If nothing else, this shows that a relatively small investment in woodland planting can result in a huge expansion of bird life. The number of bird species observed at Lanark continued to grow in the years after Elizabeth did her study. By 1996, the number stood at 158.

Elizabeth had been involved with other RMIT students in an earlier study of bird life at Lanark. The study was the idea of Jim Sinatra, an RMIT professor, and it had the active support of Murray Gunn, who had then been monitoring the comings and goings of birds at Lanark since 1950. The project was a labour of love for everyone concerned: each provided his or her expertise free of charge. The students in the first study group were Amanda Kimmins (who produced wonderful sketches and large-format perspectives), Jonathon Pearce, Graeme Quin and, of course, Elizabeth Jacka. Using my revegetation notebooks and bird diaries, they prepared charts and extracted relevant data to illustrate the changing land conditions and growth in bird numbers every ten years until 1996. The result was a ground-breaking paper titled 'A Farm Review 1840–1986'.

In 1987, Elizabeth took this study further with the 306-page paper that I referred to before. She was then a fourth-year landscape architecture student. Her work is both innovative and technically accurate to an exceptional degree. Indeed, I am not aware of anything else of its kind. In 1999 Elizabeth was awarded honours at RMIT for a final study, 'Birdscapes For Western District Farmlands'.

The eastern great egret was one of the water birds to make Lake Cicely its home for part of each year. I am particularly pleased about this, because this egret happens to be my favourite bird. For those not familiar with the species, I should explain that the large egret is a tall, all-white water bird with long legs and a long bill. Over the years the large egret has been a fairly regular visitor to Lake Cicely. I find its whiteness particularly appealing. It is not glossy white, like many other white birds, but has a kind of matt finish. It looks immaculate—all the more so when it is displaying its breeding plumage.

This egret is sometimes wrongly referred to as the white crane, which brings to mind yet another gem by the ecologist Aldo Leopold, one of my favourite sources of quotes. He wrote in his book *A Sand County Almanac* (1949): 'And so they live and have their being—these cranes—not in the constricted present, but in the wider reaches of evolutionary time. Their annual return is the ticking of the geological clock . . . [and] the sadness discernible in some marshes arises, perhaps, from their once having harboured cranes. Now they stand humbled, adrift in history.' (Leopold was probably referring here to the whooping crane, now recovering from near extinction.)

Birds were not the only type of wildlife to take advantage of the lake-side habitat we were creating. Eastern swamp rats moved in, too. As I was to learn from experience, this nearly always happens when you restore an ecological balance to what had been a degraded area: various types of wildlife, suddenly finding a healthy habitat to their liking, move in and, for a while, proliferate. This happened with the eastern swamp rats. Suddenly, the place teemed with them. In no time they had stripped Cicely's garden. They even chewed the rind off the lemons growing on a tree. The new habitat was not the only reason they multiplied so rapidly—they also lacked predators.

We had recently gone to a lot of trouble to rid the property of foxes and feral cats (David shot them with a spotlight at night). I did this for the sake of the eastern barred bandicoots that we had introduced there, but it was the swamp rats that seemed to benefit most.

Lake Cicely had been, for us, a big, expensive undertaking, but both Cicely and I soon decided that it had been worthwhile. For a time, though, it seemed we were about the only people in the district who thought so. The reaction of friends and neighbours who saw the lake was generally negative. Most were too polite to be openly critical, although you could tell that they thought the whole enterprise was a folly. Others did not hold back. They told us plainly that they thought the construction of the lake was about the dumbest thing they had ever heard of a farmer doing. What a waste of time and money! What a waste of good grazing land!

Although we were convinced that we were right and they were wrong, the negative reaction was disconcerting to say the least. I think both Cicely and I needed some reassurance at that time and, happily for us, the famous ornithologist Graham Pizzey came along to provide it. I had long known of Pizzey from his numerous books, one of which was a well-known field guide to Australian birds. While still a young man, Pizzey had given up an opportunity to take over the family leather business and, instead, devoted the rest of his life to a field of study in which he was passionately interested—bird life.

Graham Pizzey first visited Lanark in the early 1960s. He was staying with a neighbour who brought him to the property to see the old swamp we had reinstated. He spent perhaps fifteen minutes walking around the swamp by himself, studying it from every angle. Cicely and I stood there, awaiting his verdict with some trepidation. Finally, he walked over to us. 'John,' he said, 'what you've done here is wonderful.' They were his exact words. I remember them, because they resonated in my mind for a long time afterwards.

Later, Graham and his wife, Sue, came to live at nearby Dunkeld, and not long after we finished Lake Cicely he came to inspect it. His judgment was that we had done a great service to the environment. Pizzey told me one other thing. He predicted that riparian plants would return to our restored wetland of their own accord, just as the birds had done, because waterfowl would bring in the seed on their feet and in their gut. When seasonal conditions were right, he said, the plants would reappear. This proved to be true: Lake Cicely soon had a full suite of riparian plants, none of which we had planted ourselves. All we had done was fence the area off. We had a very different result in the dryland areas on the property we fenced off. There, no regeneration occurred whatsoever.

Lake Cicely proved to be a habitat for frogs as well as birds. In 2001 Damien Cook of Wetland Ecosystems Australia did a study of the various kinds of frogs living in and around Lake Cicely and was able to identify the following: striped marsh frog, spotted marsh frog, eastern banjo frog, common froglet, Ewing's tree frog and growling grass frog.

How are birds faring at Lanark today? I regret to say that over the past decade or more—say, since the mid-1990s—I have noticed that some species have disappeared altogether and that bird numbers generally have declined. I can only conclude that this is because farmers are continuing to clear bushland and drain swamps and other wetlands. I fear that a prediction made by the noted ornithologist Professor Harry Recher in a 1999 paper might still prove true. The paper, titled 'The State Of Australia's Avifauna', was published in the *Australian Zoologist* in June that year. Professor Recher wrote: 'I expect less than half of Australia's terrestrial bird species will survive the next 100 years. If I am wrong it will be because Australians modify their behaviour and change the way they manage the land, water and natural resources.'

17

FARM FORESTRY

By the end of the 1960s I had been planting trees in sizeable numbers at Lanark for the best part of ten years, so the appearance of the place had begun to change—at least in certain areas. I did not keep count, but I probably planted around 1500 trees a year on average up to 1967, the year of the famous drought, so there were now maybe 15 000 trees growing on the property that had not been there before. The trees were radiata pines (*Pinus radiata*), various eucalypts and a mix of indigenous habitat trees and shrubs. I was still following no set plan: I planted trees where and when I felt my sheep and cattle needed shelter and shade. I had not forgotten, though, my visit a decade earlier to the farm in New Zealand where Ian Young had begun to dabble in farm forestry. This had seemed a good idea then, and it seemed a better one now, because I had heard through my contacts in New Zealand of the progress farm foresters had been making there.

On my return from that visit I had taken out a subscription to the magazine *New Zealand Farmer.* The articles that interested me most, and which I looked forward eagerly to reading in each issue, were written by a farmer named Neil Barr, a pioneer in farm forestry in New Zealand, and the founding president of the New Zealand Farm Forestry Association. His articles were full of practical information about farm forestry, and I was enthused by them. I could not have guessed then that Barr would visit Lanark one day to inspect my trees. In fact, in 1989 he conducted a field day there on farm forestry. It was during that day that he performed the first culling of a Lanark tree. Until then, I had not poisoned or chopped down a single tree on the place. Ever since I have always said that Neil Barr fired the first shot in anger at a tree at Lanark.

Inspired by Barr's articles, I planted what was really a farm forestry woodlot as far back as 1961. These were the radiata pines that I planted both as a windbreak and as a screen to shield our house from the sight of the steam trains that passed along the edge of the property. Altogether, I planted 640 pines in five rows, leaving 2.4 metres between the rows and 2.7 metres between the trees in each row, the spacing recommended at the time by the Victorian Forests Commission. Although I surrounded the plantation with a netting fence to keep out rabbits and went to some trouble to prepare the ground properly, more than a third of the trees died in the first year. In 1962, I planted 250 replacement trees.

The purpose of the planting was not merely to screen the steam trains. I was also keen to see how well radiata pines, then the most common species of timber tree planted in Australia, would grow at Lanark. Knowing what farmers in New Zealand had already achieved, I was confident that growing radiatas for timber could become a profitable enterprise at Lanark.

The early results were promising and I would have liked to keep going, but establishing pine plantations is an expensive undertaking and at that time I did not have the money to do it.

I approached my bank manager in Hamilton to increase my overdraft, explaining I wanted to plant timber trees at Lanark that would generate an income for the farm down the track, but he hardly seemed to know what I was talking about. I then approached the Victorian Forests Commission, which had recently begun offering long-term, low-interest loans to finance forestry plantings, but it said no, too. It did agree a few years later, in 1968, to provide a loan for a 100-acre plantation of radiata pines that I wanted to establish at the property I had at Mount Richmond, not far from Portland, but that was because 100 acres was judged to be a genuine forestry-size block. It seemed the commission's loans were designed then to fund fairly large-scale forestry projects, not the relatively modest, multi-purpose plantings I had in mind—shelter belts consisting of maybe five rows of high-pruned pines.

The truth is that at that time the commission simply did not understand the concept of agroforestry. It certainly did not understand the concept of combining forestry plantings with shelter belts, which was then new in Australia. Moreover, according to the generally accepted view at the time, the kind of soil we had at Lanark—heavy clay close to the surface—was not ideal for growing radiatas.

The commission also turned down an application I made for a loan to grow clear-wood hardwoods—specifically, river she-oaks (*Casuarina cunninghamiana*), sugar gums and spotted gums (*Corymbia maculata*)—on a long-rotation basis. The commission said it could not be satisfied that the area was suitable for trees of this kind. Meanwhile, I continued to gather information to demonstrate that the area was completely suitable. I have often thought that if that forestry loans scheme had survived—and if it had been made freely available to farmers like me who wanted to do farm forestry plantings in shelter belts and the like—Australia's rural landscape might look very different today. More importantly, graziers' incomes might be augmented in a long-term, sustainable way.

All things considered, it is no wonder that farm forestry took a long time in Australia to get off the ground. Even the inheritance laws were against it. After the second *Landline* program went to air in 2006, someone posted the following comment on an ABC online forum. 'My father wanted to plant trees to mill back in the 1950s,' the writer said. 'He was advised not to do so as they would be valued for probate should he pass away. I am told that timber caused some people to be bankrupted in the Ravenshoe area because they planted trees.' If this were true, which I assume it was, death duties must have been a huge disincentive to a farmer interested in growing timber, although it was something I was not aware of at the time.

So for the time being my farm forestry plans had to be put on hold. My interest in the subject was revived by a trip Cicely and I made to New Zealand in 1978. We stayed for a few days at Ian Young's farm, and while we were there Ian suggested we visit a neighbouring property where the farmer, Peter Smail, had been doing big things in farm forestry.

I found it interesting that Smail was not growing his pines in plantations but, rather, was incorporating them in his working farm. He was planting wide rows of them as shelter and shade for stock, although he was also making sure they would become high-quality timber eventually by pruning them carefully. Smail later came to be regarded as a leader in New Zealand's farm forestry industry. He may not have been the first there to recognise the dual potential that tree planting offered—a more productive farm in the short term, a valuable timber harvest in the long term—but he was one of the first to try to make that potential a reality. I liked Smail. He was a busy man, and physically strong. He must also have been extremely determined, as pioneers in every field of endeavour need to be, because he had to overcome quite a few early obstacles.

I returned to Australia fired up about the potential that farm forestry offered, but my enthusiasm was rather dampened the following year, 1979, when the Victorian Forests Commission informed me in writing that it had decided farm forestry was simply not viable in our part of the Western District. Nevertheless, I continued to apply to the commission for a forestry loan. Indeed, I lodged an application just about every year, but, one by one, they were all rejected. I was told that I did not qualify.

At that time I was starting to become interested in pine plantations for another reason: I could see a lot of potential in using early pine thinnings to make treated-pine posts, then a fairly new product. It seemed to me that this would provide some income while the trees were maturing. I was one of the first farmers to use treated-pine posts in our area, as fence posts at Lanark instead of hardwood. My interest in treated-pine was to a large extent environmentally motivated. I could see that if enough Australian farmers began using them instead of hardwood posts it would significantly reduce the demand for hardwood and thus the need to log our hardwood forests. My thinking here was realistic rather than moralistic. I simply felt we ought to stop cutting down our native forests since there were so few of them left.

In 1984, finally, the Victorian Forests Commission decided that I did deserve some help with my farm forestry plans. I wanted to create a shelter belt of pines, and the commission agreed to provide me with the pine seedlings and to help with the planting. It decided to become involved after doing another inspection at Lanark. In fact, I suspect that it relented in the end not because it had really changed its mind about Lanark's suitability for forestry but, rather, because I had been badgering the people there for so long. Looking back on it now, it occurs to me that by doing this, by incessantly knocking on doors, I did, in a small way, help pioneer the idea of farm forestry, at least in my part of Australia. Certainly, very few other people seemed interested in the idea at the time.

The pines were planted in early spring 1985. To some extent, it was a trial exercise. The commission wanted to compare the growth of pines planted as seedlings with those raised from cuttings, so it planted them in four sections in a row about a kilometre long. Two sections were planted: one with cuttings and the other with seedlings from a couple of sources. This row of pines formed half of a 3.5-kilometre shelter belt along Lanark's southeast boundary. We later planted the other half with hardwood trees at our own expense.

The commission sent a small team of forestry students to help plant the pines, one of whom, a girl aged about twenty, left me with an enduring memory. The students were each issued with a forestry spade, which had a narrow blade specially shaped to create a V-shaped hole for each seedling. Now, planting trees by the hundreds can be tedious as well as tiring; this girl probably found it to be both. Whatever the reason, she fell asleep standing up and leaning on her spade. I was keeping the planters supplied with trees, and as I looked along the row I saw her. She had both hands on the handle and one foot on the blade, and she was apparently in a deep sleep. I suspect she had never planted trees before.

After the 3.5-kilometre shelter belt was finished in 1988, we erected another fence along its entire length and planted a single row of trees and shrubs to provide both habitat and low shelter. They included indigenous melaleucas, tea-trees and drooping she-oaks. My regret now is that we did not plant two or three rows instead of one. As farmers often do, I erred on the side of trying to 'save' grazing land. Today, the habitat is permanently fenced off, but sheep are allowed to graze beneath the maturing agroforestry trees. This is standard practice in wide-spaced, high-pruned agroforestry plantations: fences protecting the trees are removed, and stock allowed in, as soon as the trees are too big to be damaged by them.

18

MORE THOUGHTS ON FARM FORESTRY

Starting in 1991, a distinguished scientist from the New Zealand Forest Research Institute, Jeff Tombleson, established permanent sample plots along our southeast boundary and measured the progress of the pines growing there several times in the years that followed. Early findings tended to support the idea of growing pines commercially on a rotational basis, which makes it a feasible undertaking for an individual farmer.

Here is how a rotational program can work in practice. The farmer sets out to plant a relatively small area of pines—say, four acres (about 1.5 hectares)—each year, year after year. All things being equal, a farmer ought to be able to plant an area of this size with his own labour. The planted area would be protected with a temporary fence. After five years, pines are usually big enough to be stock-resistant, so the temporary fence that protected them can be pulled down and used to protect the next four-acre planting. The farmer would prune the trees each year (again, given the relatively small areas involved, the farmer should be able to do this himself) and after about fifteen years, when thinning of the pines begins, the farmer would start to earn an annual income from his pines, possibly from the treated-pine logs market. After 25 to 35 years, the time it usually takes for pines to mature, the farmer would then have four acres of pines available for harvest each year. Of course, he would not be obliged to harvest the four acres every year. If the market for pines was down one year, he might well choose to let the trees grow there for another year.

Over nearly the whole of the growing period, remember, the farmer's pines will have provided shelter for his stock and habitat for wildlife. So much the better if the farmer plants indigenous habitat trees with the pines, as we did at Lanark.

Which brings us to an important question: should you prune trees growing in shelter belts that are to be harvested eventually for timber? On one of his first visits to Lanark, Jeff Tombleson was amazed to see that I had not pruned many of the timber species that were growing in our shelter belts. 'Surely someone who's planted as many trees as you have would know that every potential sawlog tree needs to be pruned,' he said. I explained that I was well aware of the benefits of pruning but that I was taking part in a comparative study on pruned and unpruned trees, organised by the Victorian Department of Agriculture. 'There's no doubt about you Aussies,' Tombleson said, shaking his head. He obviously believed that the question of whether or not to prune had been answered so conclusively long before that any further examination of the subject in our situation was a waste of time and money.

Still, there is an obvious downside to pruning trees in a shelter belt, for, once pruned, it will no longer be so effective as a windbreak. After all, pruning is generally done to a height of six metres—and sometimes higher—leaving only tree trunks below that to break the wind. The simple solution to the problem is to plant a row of shrubby trees to block the wind up to pruning height. We made a standard practice of this at Lanark. In each shelter belt we would plant at least three rows of timber trees, usually pines, and one row of shrubby natives. The shelter component needs to be fenced off permanently. This is to protect the lower foliage from browsing sheep and so optimise the shelter and habitat value.

So even if the pines are high-pruned and widely spaced in rows, they can still form part of a shelter belt if there are several rows of them (I recommend five at least), and if there are two or more outer rows of preferably indigenous shrubs and trees, that will provide an effective windbreak. Naturally, these outer, shrubby rows will have to be fenced off from stock permanently.

When planting pines I sometimes mixed them with black wattles. I found the pines grew better in the presence of the wattles, which is really not so surprising. After all, plants always grow better in the presence of other plants than they do in a monoculture. Moreover, wattles add to the habitat value and they also fix nitrogen from the air, which then becomes available to the timber crop. Some foresters visiting Lanark were appalled when they saw wattles growing among the pines. Why did they react that way? The only reason I could see was that the wattles made the plantation look messy. I have found that foresters—and engineers, too—like things to be tidy.

Not all the pines we planted at Lanark will be felled for timber. The 600-odd pines I planted back in 1961—the Other Side Pines we call them, because they are on the other side of the swamp—are destined to live out their lives as trees. Although we planted them to Victorian Forests Commission specifications, we did not prune or thin them because we were told by the commission they would never be worth much as timber. This at least has proved to be true: they are not worth much as timber, but only because we failed to prune them. If we had pruned them, they would be valuable. They are huge trees now, and my son David said not long ago that he was thinking of having them cut down in the hope their timber might still be worth something. I argued against this, saying that whatever he earned from them could not possibly compensate for the resultant loss of shelter and habitat. Magpies nest in the top branches, and there is a resident whistling eagle. Moreover, the trees are a plus for the landscape. So they are still there.

I predict that farmers will soon aim to produce high-quality pruned logs. Instead of planting, say, 1650 radiata pine trees to the hectare, the normal planting density on radiata plantations today, farmers will plant perhaps 600 superior-quality trees to the hectare. The wider spacing will make it easier for farmers to prune and otherwise maintain the trees, and it will, for instance, allow them to thin inferior trees. In effect, farmers will plant only as many trees as they can look after properly, thus ensuring that the trees produce high-quality clear wood. Trees grown with a lot of space between them are likely to be big, too—much bigger than radiatas grown in today's dense plantation configuration. So while there may be fewer trees, each tree will be bigger, free of knots and, therefore, worth more when harvested.

Most people, when they invest their money, expect to see a return on it within a few years. Farmers who decide to go into farm forestry, however, need to be patient. True, there are a few tree species that will produce a financial return within a decade or two. Blue gums (*Eucalytus globulus*), for instance, can be harvested in as little as ten years if growing conditions are just right. My own view, though, is that farmers thinking of planting trees for timber on their properties should not be locked into the idea of a quick return. It would be better both for them and their farms if they took a longer view. They should even be prepared to plant trees that will not be harvested in their lifetime—and perhaps not in their children's lifetime. Oaks, for instance.

As an experiment we planted 650 'old-world' oaks (European *quercus* species) in a fenced-off five-acre (two-hectare) area, and about 80 per cent of them have survived. Most of those that died were victims of the eastern swamp rats that reappeared at Lanark once the wetlands were reinstated. We planted the oaks in the same dense configuration in which the French replanted their oak forests after World War I—five metres between rows and three metres between the trees in each row. Oaks in Australia have a rotation period (the number of years a tree takes to grow to harvestable size) of perhaps 125 years. Clearly, they offer no prospect of a return as harvested logs within one's lifetime, yet they are one of the specialty timber trees that we can grow successfully. As John Mortimer, a former president of the New Zealand Farm Forestry Association, wrote in the *New Zealand Tree Grower* in 1984: 'For too long it has been an anathema for planters to consider establishing a tree that cannot be planted and felled during a person's working life. We must wean people away from this concept and encourage the planting of longer rotation hardwoods and not worry if they come into profit during our children's lifetime or even our grandchildren's.'

Even though the oaks you plant today will earn little as sawn timber in your lifetime, they do provide other benefits. Oaks are hosts for truffles; they are great trees for children to climb; they are an excellent source of wood for those who like to whittle; and, after 50 years or so, they provide elbows for furniture making. Oaks offer one other opportunity: they can be combined with a small pig enterprise. Acorns are excellent pig feed—a fact well known to European farmers.

In 1994, the ecologist Neville Bonney wrote a paper about our farm forestry enterprise at Lanark. Near the end of the paper he had this to say: 'What Australia needs right now, when there is to be a world shortage of timber, is for banks and lending institutions to recognise this to allow loans for carefully planned farm forestry developments and for governments to offer incentives for the valuable export possibilities.' This was true then and is still true today.

19

'YOU'RE OFF YOUR BLOODY ROCKER'

Trees are slow to grow, but once they do grow they are big enough to make a difference. At Lanark, it took about 25 years for the trees I had begun planting in the late 1950s to transform the place's appearance. By the early 1980s, this once barren and desolate property was starting to look lush with vegetation. Many of the wattles had already matured and died, and the rotting wood was helping to create the balanced ecosystem that I had always hoped to achieve. Yet many farmers in the district still regarded me as eccentric or worse, and in a way that was understandable. For years, various government authorities had been encouraging and financing farmers to clear their land of trees and to drain swamps. Yet here I was doing exactly the opposite—planting tens of thousands of trees and restoring wetlands—and I was doing it all at my own expense. Who could blame other farmers for thinking I was mad?

Even the trustee company which ran my father's estate had misgivings about what I was up to. Recently, I came upon a record that I made of a conversation I had with one of the company's senior people in 1959. 'We want you to drain that 40-acre [sixteen-hectare] remnant swamp,' he said, to which I replied, 'I have just made it bigger—I have returned it to its original dimensions.' He said, 'You what? I didn't see any account for that.' I told him there was no account because I had done the work myself. He then repeated, 'I want you to drain it.' I asked, 'What for?' He then made it clear what he was really on about. 'Because we can get a government subsidy to do the work,' he said. 'Well,' I replied, 'that's no bloody reason to drain it.' As far as I can tell from my notes, that ended the conversation. The swamp did not get drained.

Cicely and I had completed the reinstatement of the wetlands at Lanark by 1970. By then, a wide variety of riparian plants was growing there, thanks in large part to the visiting water birds that had brought in seed. Not long after this, the Department of Agriculture in Victoria began establishing farm management groups in various areas of the state, the aim being to get young farmers to share ideas about what they ought to be doing with their land. The group in my area used to meet in a hotel at Digby, near Branxholme. At that time I was still feeling my way environmentally. I was still running as many sheep to the acre as anyone in the district, for instance. Yet I had already come to believe strongly that Australian farmers could not continue to regard the health of the environment as something that had little if anything to do with them. So I phoned one of the people connected with the local farm management group and told him I thought the group ought to be devoting some attention to the environment. He responded by inviting me to come along one afternoon and speak to the group about it, which I agreed to do.

I brought along some slides that we had taken of water birds on the Racecourse Swamp and I showed them during my talk. I explained that since we had reinstated the wetland some locally rare species of water birds had reappeared there. I explained how the water birds brought plant seeds with them, both in their gut and on their legs, and how, thanks to this, the wetland now had a good cover of indigenous wetland plants. Indeed, the restoration of the wetland had gone so well, I said, that ibis were now nesting there, and I showed some slides of the ibis on the screen to prove it. Whereupon a farmer at the back of the audience called out, 'So what?' Everyone laughed uproariously. I realised then that I had been wasting my time.

Now, this happened a long time ago, in the early 1970s, when people were not nearly as environmentally conscious as they are today. In fact, the nesting ibis were living proof of the fact that this big environmental project of ours, the rehabilitation and protection of a wetland, had been a success—that a bird habitat had been brought back to life.

Farmers used to believe that if they did anything out of the ordinary in the interests of the environment—if they deviated even slightly from traditional farming practices—their farms' productivity would suffer and they would be out of pocket. I have no doubt this is the main reason other farmers reacted so negatively to the environmental work we did at Lanark. They assumed that it would result, at least in the short term, in a net loss financially, which, in fact, it did.

I was interested to see that Dr Rod Bird, until recently a Hamilton-based scientist of the Victorian Department for Primary Industries, picked up on this fear of lost production in an interview with Dr Martin Mulligan of RMIT University's Globalism Research Centre. (I quote from the same interview in another chapter in this book.) Referring to our tree-planting work at Lanark, Dr Bird said:

> John and Cicely [Fenton] started at a time when it definitely wasn't fashionable, and they were regarded as rather eccentric. Some of the population would still describe this kind of work as eccentric, but much of the population has now caught up with what the Fentons were trying to do . . . But, in the end, it is not getting any easier to make a living from farming. The way out for many farmers seems to be to expand, and there is a reluctance to allocate much land to what might be regarded as non-productive use, even if there are clear long-term benefits in terms of productivity.

In an earlier chapter I quoted from an article about Lanark that Prue Powell wrote for *Bushland on Farms: Do You Have a Choice?* She touched upon the reputation I had acquired in the district for being a little crazy:

> While John [Fenton] spent all his spare time and money planting and learning about trees, his neighbours looked on at 'the town boy' with great scepticism. However, when he began putting the water back into the swamps in a horseshoe around the house, they knew he was mad. Fifty-six hectares [about 138 acres] of grazing land went into wetlands. This might have been the time John's defence line began to harden. It was characterised by the phrase, 'Go to buggery.'

I have no doubt whatsoever that my reputation for being a nonconformist stood against me in a number of ways. For instance, I was never appointed to any government committees or boards. A farmer I know with a conservative outlook believed he knew why I had been overlooked. I had just been complaining to him about how I had again failed to be appointed to two bodies in which I had a keen interest—the Catchment Management Board and Greening Australia. He said, 'Wake up to yourself, John: you'll never get on any of those committees.' I asked him why. 'Because you're a greenie and a bloody stirrer,' he said. I protested, saying that, one, I was not a greenie, and two, while I did have different views from many other farmers on how farming land should be managed I did not ever seek to provoke others with those views. He persisted. 'Of course you're a stirrer, you old bastard,' he said. 'They all talk about you. You'll never get anywhere with them.'

It is an unfortunate fact in Australia that people who come up with alternative views on how to farm and manage the land invariably get derided as cranks by the old guard. This merely reflects the rather obvious fact that, taken as a whole, country people are an extremely conservative class of people. True, some farmers in Australia have tried to do things better by doing them differently, but they are a very small minority. I used to say that if you travelled around Australia you could count on finding a farmer with innovative ideas on managing the land and protecting the environment every 30 miles or so. That is quite a distance—around 50 kilometres. This is what makes it extremely hard for farmers advocating a different approach to get any support, for at every farmers' conference they attend they are likely to find 95 per cent of the others lined up against them.

In my opinion, peer group pressure had a lot to do with the generally negative response from other farmers to the environmental work we did at Lanark. The truth is that farmers are under enormous pressure to conform to what is considered normal and traditional. They are made to feel extremely uncomfortable if they try to do things differently. This is not just a case of farmers being pressured by other farmers. Farmers are pressured to conform by government bureaucrats, too—departmental agronomists and the like—and even by their bank managers. If you step out of line—if, say, you give up using superphosphate or spraying your paddocks with chemicals—you will be made to feel isolated.

Even my own children wondered if their father had gone a little dotty. My three sons used to go to school by riding their bikes to the property's front gate, where the school bus would pick them up. One afternoon I happened to be in the house when they returned from school and pulled up on their bikes outside. The twins were then probably eight years old and their brother, John, seven. I was near a window and heard their conversation. One said, 'Did you see what he's done now?' Another said, 'Yeah, he's built another dam in the front paddock.' The third, Johnnie, said, 'That makes four dams in the paddock—the silly old bugger.'

It was true there were four dams in the front paddock. The last of them was a purpose-built wildlife dam that I built in 1969 and fenced off. The fact that I had built a dam for wildlife rather than sheep was probably what caused my sons to wonder if I had lost my senses. Actually, that wildlife dam was to prove the very last dam I built at Lanark. Nearby, I intended erecting a flagpole, and I approached the well-known Melbourne firm Evan and Evans, founded in 1877 and thus the oldest flagmaker in Australia, to design a flag for the Fenton family. It included the family motto, 'Out of difficulty I arise', which might well be Lanark's motto, too. In the end, though, neither the flag (which showed a palm tree growing from a rock) nor the flagpole went up. It was the first time I can remember Cicely really putting her foot down. 'If that flag goes up,' she said, 'I'm off.' The flag remained folded up in a cupboard.

A few farmers who at first regarded what I was doing with suspicion did come to see the wisdom of it later. One of them was a good friend of ours named Teddy Youngman, a grazier who lived not far from us. Teddy owned and flew a light plane, and one day he happened to fly over Lanark while we were building the Stockyard Dam, the one that formed Lake Cicely. He said to me the next time we met, 'What the hell are you doing down there, John? What sort of a dam is that? You'll have the whole place full of water. You're off your bloody rocker.'

Teddy later had a change of mind. He often flew over Lanark, which was on his regular route to Melbourne, and as time went on and Lake Cicely started to fill with water he apparently came to like the look of the dam. A few years later he phoned me and asked would I go over to his property, Ardgarten, because he wanted to talk to me about a dam he was thinking of building there. The upshot was that he built a dam five times as big as Lake Cicely. It was not an exercise in oneupmanship, though: that was not Teddy's style. He had simply come to believe that a large body of water would be a valuable addition to his property.

More recently, I have sensed that among Australian farmers attitudes are finally starting to change. Reality has forced them to recognise that they cannot keep doing things as they have always done—that, for their own sake if not for the environment's, their approach has to change. It is not only the younger farmers who are starting to see the light, either. Some older farmers are seeing it, too—farmers who have been around long enough to know how much healthier our land used to be.

A relative, Liz Fenton, who has a property near Hamilton, has noticed the same thing. Liz has done a lot in the district to promote the idea of planting native trees, preferably those indigenous to the area, and she grows them for sale at her indigenous farm nursery. Liz was featured in a chapter of the 1999 book *Listen to the People, Listen to the Land*, in which she made this interesting observation:

> The people who are most supportive of my work are often the older farmers. They're the ones who have generally retired and reflect on what's going on, and they look back at what the land was like when they were young and just starting out. They know they put a hell of a lot of effort into clearing, and any criticism they make is about themselves. Whereas the next generation are sometimes a bit protective of their ancestors and are afraid to see them criticised, it's the older people who have seen the land change, and it's happened over such a short time.

20

A TASTE OF FAME

By the late 1960s, certainly by the early 1970s, it was fairly well known in the Hamilton district that I had embarked upon a big tree-planting program at Lanark. Not long after this, to my surprise, I found that people far beyond Hamilton had become aware of it, too. Lanark's reputation was spreading by word of mouth, presumably that of friends and supporters. I first realised this when strangers began phoning to ask if they could come and have a look at the work we had done. As the years went on, the number of visitors of this kind increased steadily, as did the number of areas from which they came. Soon we had people arriving from South Australia. Next, they started coming from New South Wales. Before long, we were being visited by people from all over Australia.

We were always happy for people to come to Lanark and look over the place unless for some reason it was a problem to have them there on a particular day. Of course, visitors were an inconvenience in one respect. I had to show them around, which meant that whenever visitors arrived I had to stop work for as long as they were there, and Cicely often found herself obliged to offer a cup of tea. But on the whole, I quite enjoyed people coming to Lanark. Most were impressed by what they saw and told us so, which always gave my spirits a lift. It was a nice change from the generally negative reaction we received from local people.

I have a special memory of a visit by a group of agricultural scientists and farmers from Chile about ten years ago. One huge man stood back from the others as they boarded the bus to leave. He had said very little all afternoon but had obviously detected my frustration at the lack of official support for our environmental work. He put his arm on my shoulder and said with apparent emotion, 'Do not be concerned, John Fenton. Very soon the world will be aware of the value of natural capital.'

I estimate Lanark has now had about 5000 visitors since we began admitting them in 1966, which I think is a pretty impressive number for a privately owned farm. Until 1982, we used to let people visit free of charge, but financial necessity has since compelled us to charge an admission fee. I could no longer afford to stop work for an hour or two whenever a visitor turned up. There were our helpers to consider, too—experts like Rod Bird and Murray Gunn, who were not only prepared to come to Lanark to speak to visitors but were willing to do it for nothing. We needed to lighten their burden, too.

At first the visitors were individuals with a personal interest in one or other aspect of the work we were doing, whether conservation, farm forestry or habitat creation. Some of them were farmers who had heard on the grapevine what we were up to. Then, starting in the late 1970s, Lanark began to attract the attention of university people. The first to visit us, in February 1979, were groups of students from Melbourne University.

Later, we had student groups visiting us regularly from the landscape architecture department at RMIT in Melbourne. Then, academics began arriving from other states and overseas. At various times we had visitors from American and Canadian universities, as well as people from other overseas organisations, such as farm forestry associations.

Most remarkably of all, I was invited to go to what was then the Soviet Union to speak to an audience of environmentalists and farmers in the Tver region of Russia, about 200 kilometres northwest of Moscow. This was in 1992, the time of Glasnost. In the previous year an elderly Russian professor with an interest in the environment, Elvin Kalinan, had stayed with us at Lanark. Before leaving he praised our environmental work by writing in our visitors' book: 'It is a wonderful result. Heroic labour made by very nice family.' Then, as he was saying goodbye on the verandah, he hugged me and, with tears in his eyes, said through his interpreter, 'John, you must come to our country and speak to us on environmental matters.' It was an invitation I was happy to accept.

I had not been to the Soviet Union before, of course, and I quickly grew very fond of the country and its people, who, it seemed to me, had a surprisingly Australian-like sense of humour. I came to realise this when I spent ten days on an old collective farm at a village named Cressna, about four hours' driving time west of Moscow. I had hardly ever drunk vodka before, but I soon acquired a taste for it. Vodka is a trap for the innocent. During my visit to the Soviet Union, I did my best to keep a diary, recording in detail what I did each day. I soon fell behind with the diary, though, and one of the Russian organisers of my visit happened to be near me one day when I was trying to catch up. I was talking aloud, wondering what I had done on the Wednesday of the previous week, when he

Rows of pruned pines in 1996—five years after they were planted. Black wattles planted in conjunction with them are visible on the left.

My son David pruning pine trees at Lanark in 1998. Pruning is essential if trees are to become valuable as timber. *Courtesy of Jim Sinatra and Phin Murphy*

These pruned pines, plus the associated habitat planting of drooping she-oaks and moonahs, form the eastern boundary of a trial farming alley at Lanark.

High-pruned pines in stately rows: Ratty's forestry block, planted in 1997. Shown here with my granddaughter Harriet in 2009.

The path in the extensive 'old-world' oak planting established in 1993. Oaks do well in our conditions, one reason being they put down deep tap roots.

Vasey Farm has a denser tree cover than surrounding properties—as shown by this aerial shot. *Courtesy of Jim Sinatra and Phin Murphy*

My son, Patrick, with a 'fallen giant'—a huge red gum blown down in a storm at Vasey Farm. *Courtesy of Jim Sinatra and Phin Murphy*

Shorn sheep at Vasey Farm seek out shelter provided by the remnant red gums, fallen logs and natural regeneration. *Courtesy of Jim Sinatra and Phin Murphy*

Sheep graze in the Brolga Paddock. Whichever direction you look today, trees line the horizon at Lanark. Fifty years ago horizons were scarcely broken by a single tree.

The Other Side Reserve—a 65-acre area which we fenced off and planted with indigenous trees and shrubs to create a wildlife habitat.

The swimming hole that I dug for our children in the Other Side Reserve.

Our oldest grandchildren, Clive and Anna Fairbairn-Calvert (under Granny's instruction), watching a long-necked tortoise laying eggs near the Other Side in 1997.

Magpie geese at the Other Side Swamp. These birds are abundant in northern Australia's wetlands. They were once widespread in southeastern Australia. Limited reintroduction has been carried out, with some success.

The view from the front garden of Lanark's homestead. The trees visible on the far side of the lake are manna gums planted at Graeme's suggestion and, accordingly, are named the Graeme Gunn Plantation. Graeme said the trees would create an optical effect, making the lake look closer to the house, and this proved to be true.

An 'alley' between two long rows of trees.

She-oaks in the wildlife dam in the front paddock.

A view across Lake Cicely.

In my regalia after receiving an honorary doctorate in landscape architecture at RMIT University in 1997. The others in the picture are, from left: Robin Peddie (my sister), Patrick Fenton, Amanda Fairbairn-Calvert, David Fenton, Cath Fenton, Johnnie Fenton, Cicely Fenton and Charles Fairbairn-Calvert. Absent: Bronnie Fenton.

Life after Lanark: Cicely and I at about the time we left the property and moved to Hamilton.

spoke up. 'John,' he said, 'you did not have that Wednesday.' It was true: a whole day had gone missing. I had drunk so much vodka with my Russian hosts on the Tuesday evening that I was off the air for the whole of Wednesday. It was a lesson, painfully learned.

Five years after the Russian trip, our environmental work was brought to national attention in Australia by the ABC's *Landline* TV program. It was by no means the first media exposure we had received, however. A journalist working for Melbourne's *Age* newspaper, Rod Usher, wrote an article about us in the late 1980s that appeared to create a lot of interest. This was really the first occasion Lanark had made the metropolitan news. Then, in the mid-1990s, another journalist, Richard Cornish, wrote at length about Lanark, Cicely and me in the *Sunday Age*. Under the heading 'The arbour of love' was a photo of Cicely and me walking hand in hand beside Lake Cicely. Cornish's article must have been widely read, because we had plenty of feedback from it later. This article probably got the *Landline* people interested, because they approached us not long after it appeared.

A *Landline* film crew spent three days with us at Lanark, filming the place and interviewing the family, which, for the ABC, amounted to quite a large investment of time and effort. They even brought a helicopter with them. The presenter of the program was Pip Courtney, a Tasmanian-bred journalist whom we came to like personally and admire professionally. In fact, I was extremely impressed by the professionalism of the entire crew. The program went to air in 1997 and drew a big response from viewers. Clearly, the kind of environmental work we were doing at Lanark struck a chord with the public. The show proved so popular that Pip returned with a *Landline* team in 2006 to do another program. By then, of course, my son David was Lanark's resident owner, and the program focused in part on the family's attempts to make the farm more productive while still caring for the environment.

I found the response to the *Landline* programs, especially the first one, very satisfying. The mere fact that so many people were interested in what we had done was further confirmation for me that it had all been worthwhile. I was also pleased that the help and encouragement certain local people had given us over the years—people like Rod Bird and Murray Gunn—was publicly acknowledged. The plain fact is that, without the scientific expertise and moral support of people like these, Lanark as we know it today would never have materialised.

What I did find significant about the initial reaction to the *Landline* program was that the people who took most interest in the show, as measured by the numbers of letters and phone calls we received, were not traditional farmers but people new to farming, including city people who had bought rural properties. This was not really so surprising. After all, it is always harder to accept new concepts if you have been adhering to old practices all your working life. I also think that people new to farming see things with fresh eyes, which means they are better able to recognise that our landscape has serious problems. Moreover, if they have an off-farm income they can better afford to do something about it.

Since that first *Landline* program, though, I have noticed a growing interest among farmers generally in the environmental work we have done at Lanark. I find that farmers who were entirely indifferent to it ten years ago often come up to me now to speak about it. Perhaps the message is finally getting through.

One consequence of the publicity was that I was asked more often to speak at seminars and conferences, some of which were attended by eminent scientists. Now, public speaking did not come easily to me. In fact, I was once told bluntly that I was no good at it. In 1981 I had applied for a Churchill Fellowship, saying I wanted to further my study of 'the integration of farm management plans that will embrace the conservation of soil and water [and] revegetation of tree and shrub cover in such a way as to optimise the indigenous wildlife on the farm'. I submitted that this would result in a sustainable farming system offering farmers a future of stable financial returns. The system, I added, would incorporate a small but high-quality forestry enterprise.

I had to make my case in person at the final interview, where I faced a panel of three bureaucrats, chaired by a senior executive in the Victorian Forests Commission. After leaving the room to consider my application, they returned to say it had been refused. The chairman of the panel added for good measure that I had made a 'poor personal presentation' and that I did not speak well or clearly. He may well have been right about my presentation and my speech, but I suspect that he and the other panellists, like most people in the forestry and agricultural sectors 30 years ago, had little or no understanding of the kind of farming system I was proposing. Anyone advocating this type of approach was liable to be brushed aside as a greenie or as someone wanting to indulge in an alternative lifestyle.

Still, I persisted as a public speaker and gradually learned to give a reasonable performance. In March 2000 Cicely and I found ourselves sharing a stage in Melbourne with the environmentalist David Suzuki and speaking to 1200 people. The occasion was an International Landcare conference titled *Changing Landscapes: Shaping Futures.* (One of the other speakers was Pat Dodson, the Aboriginal leader, who as a boy was sent from his home in Broome, Western Australia, to board at a Catholic school in Hamilton, Monivae College. He ended up becoming the school's head prefect.) Speaking at the conference was, for Cicely and me, an unnerving experience. I have always felt stressed when speaking to large audiences. I used to hope that, with practice, I would get used to it, but this has not happened. I did not get to meet Dr Suzuki, unfortunately. I think he slipped away as soon as he had finished speaking.

On the day of the conference, Cicely and I were both on the stage. I had been invited to speak at the plenary session, and I began by saying that Cicely and I had never addressed an audience together before. I then added, 'This could well be a Punch and Judy show.' It was not much of a joke, but it seemed to put the audience in a good humour, after which the presentation went down well. The audience did not even seem to mind that, unlike all the other speakers, I illustrated my talk not with a PowerPoint program but with around twenty old-fashioned slides, the first of which had been taken back in 1955.

I did work out early on that I should simply be myself when I spoke to an audience, regardless of how erudite the audience happened to be. In other words, I should never pretend to be anything other than a farmer with a passion for restoring and protecting the environment. I remember speaking at one fairly high-powered environmental conference in Adelaide in 2002. After a few other speakers and I had said our bit, questions were invited from the audience, whereupon someone got up and said, 'I'd be interested to know what John Fenton thinks of EBMPs.' After a long pause I replied, 'I'm sorry, but I've never heard of EBMPs. I haven't the faintest idea what you're talking about.'

A few people, probably farmers, started to clap, but most of the audience greeted what I said with a scornful silence, which made me bristle. I went on:

> That's one of the problems with you people—setting agendas for farmers to follow. You think that you're talking to farmers, but you're talking to them in a foreign voice. We farmers know how to plant trees and we know all about soil. We know about the environment and we know it from first hand. If you want to converse with us about the environment, you need to do it in a language both of us can understand.

I did learn later, incidentally, that EBMP stood for environmental best management practices, although I am still not completely sure what that means.

21

A CORPORATE VENTURE

In 1981, when we were weighed down heavily by the cost of our tree planting and other environmental projects at Lanark, a bright idea came to me. While we waited for our farm forestry plantings to produce a return, I might be able to ease the financial pressure by marketing the expertise that I had gained over the previous 25 years—that is, expertise at siting, planting, protecting and nurturing farm trees. In other words, I should start a tree-planting business. I thought about it carefully and decided to do it—although not on my own. I went into the venture with a partner, Roger Young, a son of Ian Young, the farmer I had worked for in New Zealand in the summer of 1957–58.

I described in an earlier chapter how I ended up working for Ian quite by chance. As a result of that connection, our families, the Youngs and the Fentons, became entwined in a quite remarkable way. The fact that the two families were about half a generation apart (in age, I was roughly halfway between

Ian and his children, and his children were roughly halfway between me and my children) probably had a lot to do with this. Just as I had worked for Ian, Ian's son, Jim, came to work for me at Lanark. As I did at the Youngs' farm, Jim became one of the family. He babysat our children when the need arose, which I had done in New Zealand for him and his brothers. There were further exchanges. Jim's brother, Roger, came to work at Lanark, and in due course, my daughter, Amanda, went to New Zealand and lived and worked at the Youngs' farm, which, by now, Roger was running. Amanda carried on the inter-family tradition by looking after the children of Roger and his wife, Lois.

Our families later became entwined in another way. Some years after his original stint at Lanark, Roger moved to Australia and bought a farm at Condah, about fifteen kilometres from Lanark. He decided to make the move after his father sold the family farm in New Zealand. Roger's new farm probably did not meet all his expectations, for the country here was not nearly as good as the Canterbury Plains, but I think he did pretty well. He established a deer farm, the first in the district, and he also ran sheep and did some cropping. Naturally, we saw quite a bit of each other.

For several years before Roger arrived, I had been doing some outside work as a tree-planting consultant. For a while I worked with a leading landscape architecture and planning firm called Tract Consultants, helping it with broad-acre projects. In fact, I probably learned more from the firm's highly skilled staff than they learned from me. After that, Portland Council appointed me to implement a large-scale, rather complex landscape plan for the new Portland Airport, which Tract Consultants had designed. The airport job was a big one, and I felt I needed help. So I approached Roger and asked if he would be interested in joining me in a tree-planting consultancy. He liked the idea, and soon after that, in 1982, we formed a company to conduct the business, which, at Cicely's suggestion, we named Rural Trees Australia.

Roger was about twenty years younger than me, but we worked well together and the business prospered. We were joined soon afterwards by Murray Gunn. By now it had become clear to us that farm planning was not just about practicalities; it was also about aesthetics. A farm had to look good as well as function well, which, in turn, meant we needed the skills of a landscape architect. Professor Jim Sinatra of RMIT University was willing to provide advice whenever I asked for it, but what our business really needed was a small staff of landscape architects preparing plans on site. Whereas we had previously based ourselves in The Hut at Lanark, we now set up office in a house we built in Branxholme village, and recruited three graduates from Sinatra's landscape architecture faculty to occupy it. Later still, we sold the house at Branxholme and moved the office to an old sawmill complex, complete with office and sheds, in Hamilton. We felt we needed to have a profile in what is, after all, the biggest rural centre in Victoria west of Ballarat.

I also enlisted the services of an old friend, Bill Sharp, a fellow Branxholme farmer who had worked for many years for the Victorian Soil Conservation Authority. He brought with him the skills needed to handle important matters such as land-use practice, soil degradation, streamside problems and farm water-supply design. Finally, we formed a permanent works team of three people, headed by Frank Carlin under the control of Roger Young. By the end of our third year, we had a full-time workforce of ten people, including myself and

Roger. Our modest enterprise had suddenly become a sizeable business.

For a while, our firm struggled to cope with the demand from property owners, most of them corporate people or hobby farmers, who hired us to design tree-planting layouts for their land, and then to fence off and plant the trees. (We employed my son John and a fencing specialist named Graeme Bowen to do the fencing.) Our clients were not just in Victoria, either. We also worked on numerous properties in New South Wales and South Australia.

I remember one client in particular, a Sydney man who owned a big property in the central west of New South Wales. He gave us a virtual carte blanche to revegetate the entire property with trees according to a plan of our own design. To ensure that the trees we planted were suited to local conditions, we decided to grow them from seed collected nearby. Unfortunately, there was nobody in the area capable of growing the trees to our specifications, so we had to grow them ourselves. Murray did a botanical survey, collected seed from the nearest indigenous source, grew the trees in paper tubes in Victoria and trucked as many as 35 000 of them to the property, and then machine-planted them in accord with our detailed farm plan.

I am sometimes asked why I did not make a regular practice of growing my own trees from seed. The reason, frankly, is that I was always so busy planting them that I did not have enough spare time to grow them. Growing trees from seed is, after all, a specialist skill—and a time-consuming one at that. Moreover, trees were not expensive if you bought them wholesale in large numbers as tube stock. On the contrary: I always thought trees were surprisingly cheap. They still are.

Provided you do it in bulk, you can still buy tube stock at barely a dollar each.

Neville Bonney, who had the first indigenous plant nursery in our area, was a valuable source of both trees and advice. He was always willing to share with us his immense knowledge of all things natural, and he supplied us with many thousands of specially grown trees. In 1983 he invited us to help with an interesting project he was working on in the southeast of South Australia. The plan was to establish a commercial crop of banksias to supply cut flowers, mainly for export to Europe, which meant planting banksias in very large numbers. With the help of a tree-planting machine, a team of three—Cicely, Roger and Murray—managed to plant 14 000 banksias in four days.

In 1985 the Ian Potter Foundation funded a 'Whole Farm Planning' program to foster sustainable farming, an important part of which, naturally enough, was planting farm trees. We tried to interest people organising the program in our tree-planting methods but had no success. I remember one of them telling me one day at a symposium at the Melbourne Showgrounds that the machine-planting system we had developed was a waste of time. It would never work, he told me in front of a number of other people, adding for good measure, 'You're dreaming.' The funny thing was that people connected with the Ian Potter Foundation program later approached us wanting to borrow this same tree-planting equipment.

Much of the planting we did was by direct seeding, a method that had been found to work efficiently as far back as the 1930s, when it was used on the basalt tablelands in Victoria, mainly to plant sugar gums. But conditions had changed since then, largely because of soil compaction and the use of artificial fertilizers on exotic pastures. In the late 1980s there was a push to develop new direct-seeding techniques. Roger

Young made a significant contribution here. He designed and built a precision-seeding machine that was able to deposit small quantities of seed on the ground at specified intervals, usually three metres. Much of the seed we used was drawn from a seed bank that Murray Gunn had built up over time by carefully selecting species and collecting seed. Roger's machine made it possible to direct-seed forestry species.

After its extremely promising start, Rural Trees Australia eventually failed. Altogether, the business operated for about ten years. My own view is that it grew too big, too quickly: we were simply overstretched. I also think we set a standard of service that was too high to be maintained. It was simply not possible to do things as well as we did them and end up consistently earning more than we spent. Of course, if we had received some kind of government subsidy to plant the trees and aid our research, things would have been different. As I have observed elsewhere, however, governments have always been strangely reluctant to support private-enterprise attempts to improve the environment.

The late Eric Rolls, the farmer–author who wrote superbly about Australia's natural heritage, had some good things to say about Rural Trees Australia in a 1993 book, *From Forest to Sea: Australia's Changing Environment*. With considerable perception, Rolls picked up on the fact that we went to extreme lengths to get things right, which, as I suggested, may have contributed to the business's demise. Referring to the landscape architects that we employed—David Hay, Don Thomson and my niece Sarah Peddie, whose married name now is Sharp—Rolls wrote:

> The big maps they draw when planning a property are meticulous and consider what will happen over the next

> fifty years at least. Tree lines will never cut off the wind from a mill, or a new gully wash in heavy rain because water has been turned wrongly . . . They do not overlook the importance of single trees and mark them on the maps, both those in place and those to be planted. A lone big tree can transpire as much water as several in a clump. It produces more leaves and the wind can get at it better to hasten evaporation.

Rolls then went on to make an interesting point. Still referring to Rural Trees Australia, he wrote: 'This work is exciting. Farms have never been considered as entities growing over the next 100 years. Most of the planning is about next year's crop or the next wool clip.' In this, sadly, he was absolutely correct.

22

HARD TIMES

In recent chapters I have looked, one by one, at the various tree-planting and other environmental projects we undertook from the 1960s to the 1980s. Lanark changed beyond recognition during those years, and we, the Fentons, changed, too. Cicely and I grew older—and our children grew to adulthood. Fortunately, we did not have to send the children away to boarding school, so we were able to watch them growing up. All four went to the primary school at nearby Branxholme, which was then, and still is, a typically caring country school. The four were able to do their secondary schooling locally, too—at Hamilton College, an excellent Uniting Church co-ed school in Hamilton—thanks in part to the fact that a school bus ran past Lanark's front gate.

Our children grew up knowing virtually all there was to know about the business of farming. They learned it on the job, too. All four helped with work on the farm, especially during school holidays. They did all the usual tasks: fencing, marking lambs, rouseabouting at shearing time and, of course, planting trees. They also had their own day-to-day responsibilities. Johnnie provided us with daily fresh milk by looking after and milking a cow; Amanda looked after the chooks and kept us in eggs; David and Patrick both set up workshops where they met the family's needs in various ways. David learned to plait leather and made belts for us. Patrick worked in leather, too, and became the family's shoe repairer. We never paid them for the work they did: it was accepted as being part of growing up on a farm. Looking back now, it occurs to me that, collectively over the years, our four children did a huge amount of work on the place. They made a genuinely important contribution to Lanark's development.

By 1980, all our children had left school. All four went on to do farm work of one kind or another. Amanda could have gone to university after leaving school, but she decided against it in the end, preferring to work on the land. She did farm work in New Zealand, England and Scotland, and in 1987, after returning to Australia, she married Charles Fairbairn-Calvert, a grazier who also does a lot of cropping. Their property, Banongill East, is near Skipton, west of Ballarat. I am pleased to say that at Banongill East she and Charles have done many of the things we did at Lanark: reinstating wetlands, protecting remnant vegetation and planting large numbers of trees (more than 30 000 at a recent count) both as shelter belts and timber woodlots.

After leaving school Patrick went to Canada on an international rural exchange program, and he later worked with the Department of Agriculture in Hamilton. David went overseas, too, and worked on a farm in Scotland for a while, but for the most part his working life was centred on Lanark. In 1992 he took over as Lanark's manager. Johnnie tried his hand in various lines of work around Australia, all the while gathering valuable know-how and experience. Most of the work he did was in one way or another connected with the rural industry—contract harvesting on grain properties and working as a mechanic on a remote cattle station in Western Australia—although for several years he was the publican at the Beachport Hotel in Beachport, South Australia. He also went to England with a touring cricket club and stayed on there for a stint of farming. Johnnie is now running his own building business in Robe, South Australia.

A question of huge importance soon presented itself: what were we to do with Lanark in the long term? I explained earlier how my father had left the property not to me but to my children, the idea being this would save me having to pay death duties. Under the terms of his will, each of our four children was entitled to a quarter of the property. Now, it was obvious that a property of Lanark's size would be able to support only one owner, so we needed to come to some mutually acceptable agreement on how family assets, including Cicely's, would be distributed. So in the early 1990s we got the four children together to work out an estate plan. Our solicitor and accountant were there too. Today, succession planning is becoming a fairly common practice, but it was not common in the early 1990s, so in that sense we were ahead of our time. Our daughter, Amanda, was married by then, but all three boys were still single, which made the exercise a good deal less complicated.

An agreement was duly reached, under which David was to end up with Lanark, Patrick would get Vasey Farm, John would end up with a property that we bought in the Wimmera (Johnnie preferred cropping to running livestock) and Amanda would receive some money in return for her share of Lanark. Admittedly, this is a simplified summary of what was agreed, for the agreement did involve some fairly complex juggling of money and property between the four children. The end result, though, seemed fair and acceptable to all, and it meant that the question of how the estate was to be shared was sorted out long before the sharing had to take place. I commend the idea of estate planning in advance to any farming family in a similar situation.

In one sense, we were lucky there was still much of an estate to share, because for years, off and on, Cicely and I had struggled to make ends meet. We were under financial pressure of varying degrees for quite a long time—from the early 1970s, probably, through to the early 1990s. There were numerous occasions during this period when survival became a real battle—when the farm was simply not earning enough money to pay the bills. On one occasion, when funds were particularly scarce, Cicely took a job in Melbourne. That was around 1990. She based herself there for six months, working at a nursery and living at the home of the nursery's owners, who were in the US the whole time, setting up a nursery. One sometimes hears of married women who choose to go back to work after their children grow up, perhaps because they are bored with being at home or perhaps because they are seeking personal independence. This was certainly not true in Cicely's case. She did not leave the farm and take a job because she wanted to; we simply needed the money.

I once took a job in Melbourne myself, working for Tract Consultants, the firm of landscape architects and planners I spoke about in the previous chapter. Cicely and the children ran the farm between them in my absence. In the end, the job came to nothing, as did, eventually, our company Rural Trees Australia. I had hoped my earnings from Rural Trees Australia would subsidise our farming operation at Lanark and enable Cicely and me to continue our environmental work there without being stressed financially. But this was not to be.

Financial necessity also forced me to sell an 1100-acre (450-hectare) property that I would dearly have liked to hang onto. I bought it in the mid-1960s. A big part of its appeal was that it adjoined the Mount Richmond National Park, which is close to the coast, twenty minutes west of Portland. Another part of its appeal was that it included 100 acres (40 hectares) of virgin forest. I was only in my mid-thirties then, but, knowing how long trees take to mature, I was already counting the years and wondering if I would live long enough to see a mosaic of shade and shelter at Lanark. At Mount Richmond, the trees were already there.

In the end, I was obliged to sell the Mount Richmond farm. My bank manager called me into his office one day in 1981 and told me that interest rates had risen and that I had to reduce my debt by selling the property. I told him I had just planted 100 acres of pines there, plus 100 acres of lucerne. The former meant nothing to the bank manager. I am sure I was the only farmer he knew who was planting pine trees. It was no use arguing with him, so I just did what he told me and Mount Richmond was sold soon afterwards.

We did scrape by in the end, but only just. As I discovered, scraping by is not a pleasant experience. In our case, this was not because we had to go without. As a family we may have been severely restricted in where and how often we took holidays, for instance, but we never went without necessities. No, the really distressing, debilitating thing about being stretched financially is the emotional toll it takes—the stress that it imposes, day after day, night after night. I certainly suffered from stress because of it. Cicely, I know, suffered more than me.

Why were we doing it so tough? The simple answer is that we spent so much time, energy and money on our environmental work at Lanark that we did not have enough time, energy and money to spend on our farming business. We were both well aware of this at the time. Yet, even when our financial worries seemed all but overwhelming, I do not think that either of us ever regretted doing what we had done at Lanark. Of course, by the 1970s, when we first encountered hard times, the results of our environmental work were already there to be seen and enjoyed. The trees we had planted were changing the landscape. All kinds of wildlife had moved into the bushland reserves we had created. If we had run into financial trouble in the early 1960s, however, when there was little to show for the environmental work we had begun only a few years earlier, we would have found it much harder to press on. Quite possibly, we would have given up.

In the light of all this, what hope do ordinary farmers have of being able to rehabilitate their properties environmentally? Sure, planting trees is good, but how many trees does a farmer need to plant and what is the secret of getting them to grow? And how much better off will the farmer be once the trees have grown? I will set out what I have discovered about these and other relevant issues in the final third of this book, beginning in the next chapter with perhaps the most important issue of all: farmers may want to restore their properties environmentally, but can they afford to do it?

23

IS LANARK A MODEL FOR FARMERS?

People who know Lanark, with its many magnificent rows of mature trees, its wetlands, its bushland reserves and its big plantations of commercial timber, have been generous enough to say that I have succeeded in producing a sustainable rural property that farmers all over Australia should use as a model. The New Zealand agroforester Jeff Tombleson had this to say about Lanark: 'This is one of the most important properties in Australia, since it will lead the change from primary wool production to a diversified farm economy through farm forestry.' In other words, I was showing other farmers the way ahead to sustainability. It was a flattering suggestion, but I am not sure I agree with it. Let me state my position.

First, it cannot be disputed that the 100 000 or more trees I planted over a period of almost 50 years have made Lanark a far more productive and financially secure property than it would be now if I had not planted the trees. I have already explained why this is so. As long as they are planted in the right place and in the right configuration, trees protect farm animals in winter; they foster growth in the paddocks in summer by blocking wind and so reducing evaporation; they enhance the fertility of the land by creating organic matter that eventually enriches the soil; they attract birds and insects which are good for the environment and therefore good for the health of the land; they improve the capital value of the property; and, if the right species have been planted to be harvested for timber, they provide the farmer with an ongoing source of income in the longer term.

All in all, I am convinced that planting trees on farms, any farm, is virtually guaranteed to make that farm more productive and sustainable. So why am I reluctant to say that my way of managing the land and conserving the environment is the best way? Why am I reluctant to recommend to other farmers that they undertake a tree-planting program on the same scale as mine? Or set aside a sizeable proportion of their land as a wildlife habitat? Or create a wetland where bird life can flourish?

The reason has all to do with time and money. Doing the kind of environmental work I have just outlined is an extremely time-consuming and expensive operation. Frankly, I believe that only a tiny proportion of farmers—specifically, those with substantial off-farm incomes—can afford to do it comfortably. Farmers who rely for a living largely or entirely on their farms would risk running themselves ragged, both financially and emotionally, if they tried.

As I admitted before, I came close to going under myself. In retrospect, I can now say that, although the work that I did at Lanark was enormously beneficial for the environment, it was work that, as a farmer, I could not justify on a cost-benefit basis. I admit now I got the balance wrong—that is, the balance between my obligation to protect and improve the environment and my obligation to my family to run a profitable farm. My son David has done a lot to restore the balance since taking over the property. In particular, he has mastered the technique of producing high-quality fat lambs from a self-replacing flock. I had always concentrated on growing wool. Now, wool is a superb natural product that is better than the synthetic product it competes against in just about every way; that is beyond dispute. But there is not much point in growing wool if people do not want to buy it. Unfortunately, we farmers are price-takers, not price-makers.

Of course, things would be different if, in return for taking care of the environment, farmers were paid a stewardship fee, a concept I will discuss in the next chapter. Then, every farmer could afford to do as much conservation work as needs to be done. Otherwise, I frankly do not believe Lanark can be presented as a model that farmers should aim to reproduce. In most cases, reproducing what I have done at Lanark would be beyond them, both time-wise and money-wise. Still, something needs to be done, and I believe a scaled-down version of the Lanark model would be within the capacity and budget of most farmers.

Nowadays, I think everyone accepts that, taken as a whole, Australia's farming land is degraded and in need of rehabilitation. Unfortunately, the job of rehabilitating a working farm, which really means a farm of 1500 acres (600 hectares) or more, is just too big, in a practical sense, to be completed in the lifetime of the farmer. True, I managed to do it. Lanark was bare and waterless—an environmental wasteland—when I set about rehabilitating it. It took me 50 years, but I succeeded in turning it into an environmentally healthy, economically sustainable rural property, complete with wetlands and other wildlife habitats. I have to say, though, that I was an exception. I had a passion for the environmental work that I undertook. Some of my friends, I know, regarded it as an obsession, and they may well be right. Without that passion, obsession—call it what you will—there is no way I could have persevered with the work over half a century, for it certainly took its toll, physically, financially and emotionally.

Still, people who visit Lanark invariably go away impressed. They gaze with surprise and admiration at the row upon row of maturing trees that stretch into the distance, and they marvel at the wetlands and the wildlife reserves. I could not count the number of visitors who have said to me, 'Every farm ought to look like this.' That is probably true. The problem is not every farmer is capable of making it happen.

24

STEWARDSHIP: THE KEY TO THE FUTURE

As we know, the community expects farmers to take good care of whatever piece of the environment is under their control. As we also know, farmers need to make a living from their land. Now, there is clearly a fundamental gap between community expectations and the ability of farmers to meet those expectations and still make a living. Can the gap ever be closed? I am sure that it can, and I have no doubt that the best way of closing it is by adopting a system of stewardship payments to farmers. The concept is not new, of course. For some time now stewardship payments have been made to farmers in the United States, Britain and certain other European countries. In Australia, the National Farmers' Federation has been calling for the introduction of stewardship payments for several years, so far with little success.

The scheme I have in mind—which I think is roughly the same as the scheme advocated by farmers generally—would work like this. The federal or state governments would pay farmers a regular, ongoing fee for setting aside land for the environment and managing it in an agreed way. I am not referring here to ordinary farming practices that also happen to be environmentally sound, such as leaving stands of trees as shade and shelter for their stock. Leaving the stands of trees benefits the stock—and therefore the farmer—as much as it does the environment. Rather, I am referring to conservation work undertaken by farmers that produces no financial return for them. An obvious example in Lanark's case is the 65-acre (27-hectare) wildlife habitat, which we call the Other Side Reserve, that we set aside many years ago.

Of course, it might be argued that the habitat would improve the value of the property if ever my son wanted to sell it. In other words, a 65-acre wildlife reserve might seem an attraction to a potential buyer, especially one who was interested in native plants and conservation. The truth is, though, that I set aside over 5 per cent of my land for the benefit of the environment. In other words, I did it for the general good. Should farmers like me be compensated with government money for sacrificing personal gain in the interests of the general good? I am certain that they should be. This is not so much to reward the small minority of farmers who have already done their bit. Rather, it is to encourage other farmers who have made no specific contribution to the environment, in most cases because they could not afford the financial cost, to do the same.

It all comes down to this: farmers are the only people capable of conserving and improving the environment. Governments cannot do it. In a country as big as ours, governments would probably need to employ a work force of a million people to manage and maintain the environment. For the same reason, organisations like Landcare and Greening Australia cannot do it either, no matter how well intentioned they may be. The job is far too big for them. Only farmers can do it, for the simple reason that they are the people on the ground. They occupy and control the land; they own and operate machinery that would be needed for environmental work; and they sit on local councils. It has been estimated that farmers (who represent only a tiny proportion of Australia's population) control about two-thirds of the Australian land mass. Looking after the environment is a massive responsibility for such a small group of people.

Of course, if farmers were a wealthy lot with plenty of spare time and spare money on their hands, then maybe they would be able to do, perhaps out of a sense of community spirit, what needs to be done to preserve and enhance the environment. In fact, though, most farmers nowadays are doing it very tough financially. They are so hard-pressed trying to make ends meet that most of them do not come close to being able to afford the time and money required to look after the environment in the way it should be looked after. The only way they can afford to look after it properly is by being paid to do it, preferably under a stewardship scheme.

Payments like these would not be handouts. They would be payments for a service provided to the community—indeed, to the nation as a whole. I would argue that farmers who spend time and money caring for the environment are really providing a service for the world, for the environment is obviously a global concern.

It must be stressed, too, that the payments would not be for the whole of the farm. They would apply only for the areas set aside under the terms of the stewardship deal. As part of the deal, a covenant could be placed on land set aside as wildlife habitat, and this covenant would remain after the property changed hands. The covenant would be concise and include a carefully considered, long-term management plan. Would the presence of such a covenant reduce the market value of the property? I am sure it would not. On the contrary: I believe that, as time went on and people became more environmentally conscious, it would increase the value. The presence on a property of a wildlife refuge would make the property more attractive to a majority of potential buyers.

In 2008 a group from the Royal Geographic Society of South Australia visited Lanark to inspect our conservation work. One of them took me aside and explained to me that he had been engaged by a farmers' organisation to prepare a proposal for a system of stewardship payments, which, I assumed, was to be presented to the state or federal government. He asked me how much I thought our contribution to the environment at Lanark was worth. In other words, he said, what kind of stewardship payment would my son David need to be paid for continuing the kind of environmental work we had done at Lanark?

I replied, a little tongue in cheek, that a dollar an acre might be enough. He looked surprised so I explained the point that I was trying to make. I told him that if David were to be paid even one dollar an acre, a paltry sum, this would at least demonstrate to every other farmer in the district that what had been done at Lanark was worthwhile—that it was really something of value. He saw my point, but he admitted that the annual payment he had in mind was more like $100 per acre of that part of the farm permanently set aside for environmental purposes.

Stewardship payments must not be confused with farm subsidies. The two things are quite different. Stewardship fees would be a specific payment to farmers for a specific service provided—namely, caring for the environment. Farm subsidies, on the other hand, are just that: subsidies to help cover the cost of farming operations. I suspect many Australian farmers would not wish to receive subsidies of the type paid to farmers in the US, even in the highly unlikely event of the Australian Government ever being willing to pay them.

Stewardship payments have one key advantage: they would be politically acceptable. By this I mean that Australian voters as a whole, most of whom live in towns and cities, would be happy for farmers to receive stewardship fees, provided they were paid only after strict conditions were met, whereas they would probably not be happy for them to receive farm subsidies. Urban Australians still cling to the old and entirely false idea that people on the land are well off, so they would see farm subsidies as a form of welfare for the wealthy. I believe they would happily accept the idea of stewardship payments, however, because by and large urban dwellers are environmentally conscious—they have an emotional attachment to the bush. They want the countryside they drive through on holidays to look well kept and attractive—which effectively means they want Australia's rural environment restored and protected—and they would be willing to pay for it with taxes. If the Minister for the Environment were to introduce a stewardship payments scheme tomorrow, I have no doubt he would be applauded by a majority of voters in towns and cities.

As things stand now, conservation work is reckoned to be the responsibility of the farmer who owns the land in question. Let me cite a recent example in our part of Victoria. A farmer I know in the Hamilton district has a rare plant called curly sedge (*Carex tasmanica*) growing around a seasonal wetland on his property. This species has survived only in certain parts of Tasmania and Victoria, and even in these areas it is considered a threatened species. The problem is that its habitat (it grows around the edges of swamps and streams) has been shrinking, largely because of more and more incursions by grazing animals. Recently, an officer from the relevant government department came to see the farmer about protecting the curly sedge. Until then, the farmer had never heard of the plant. He had seen it growing near a creek but had no idea what it was. The department officer explained it was imperative that the curly sedge should be fenced off from stock, and he assured the farmer that the department would pay the cost of the fencing material.

It was up to the farmer, though, to erect the fence. But erecting a stock-proof fence around and through a wetland, which invariably involves the construction of a crossing and gateway for access across the stream, is an extremely time-consuming job for farmers working on their own. Protecting the curly sedge would thus have occupied many hours of the farmer's time, which he could ill afford. As it was, he was battling to keep his farm afloat financially. So he said to the department officer, 'Thanks, but I don't think I'll worry about it.' The department officer, who probably imagined he was offering the farmer a good deal by agreeing to pay for the fencing material, was clearly taken aback. 'I'm sorry, but you have to worry about it,' he said. 'This plant is endangered.' But the farmer refused.

Now, some people on hearing this story might well feel appalled at the farmer's attitude. In fact, I happen to know that the farmer in question is extremely responsible and caring when it comes to looking after the environment. He simply could not spare the week or two it might have taken to erect the fence. He was certainly not indifferent to the fate of the sedge. Far from it; he could see that the grazing regime he had been operating in that area for the previous 25 years was allowing the sedge to survive and even flourish in the wetland.

Moreover, he could not be sure it would continue to flourish if he fenced off the swamp and changed the grazing regime. I checked with the farmer just before I began this chapter, and he told me the curly sedge was still growing happily.

Yet imagine how different things might have been if a stewardship payments scheme had been operating. Then, the farmer would have been under an obligation to do all he reasonably could to protect the sedge, which may or may not have involved the erection of the fence. Between them, the farmer and department officer could have monitored the sedge's health. If after a time it became clear that the sedge had to be fenced off, then the farmer would have been obliged to do the fencing and the department to supply the fencing material. Then, if fencing the area meant his sheep were deprived of water, the department would have provided a solar-powered pump to supply water from a nearby creek to a trough in the next paddock. In other words, it would have been a cooperative effort between the community, as represented by the department officer, and the farmer. I believe this is precisely the kind of cooperation that is needed if we in Australia are to restore our environment to an acceptable state.

There are other steps that the Australian Government could take to restore and protect the environment in rural Australia. For one thing, it could approve the waiving of all rates and taxes imposed on land that farmers set aside and manage solely for the benefit of the environment. Most if not all of the charges so waived would have been imposed by state and local governments, and it would be up to the Australian Government to reimburse them for it. By definition, the land that farmers set aside for the environment would cease to be productive land. It would be easy enough to identify non-productive land of this kind from aerial photos. The same method is used to identify farm forestry areas under the Forestry Rights Act. Alternatively, the land set aside for environmental reasons could be assessed by the local government valuer, which may be an even better idea.

Finally, let me suggest an initiative that I am convinced would do wonders for the environment in Australia and would also prove popular with voters—the creation of a program along the lines of the Civilian Conservation Corps that President Franklin Roosevelt established in the US during the Depression. In 1983 I read a book about the CCC, as it came to be known, and made the following note in the margin: 'It is fifty years this month since President Roosevelt formed the CCC. I think the conditions are now much as they were in America then. How can I initiate a program like this?'

The CCC was a program that not only provided work for up to half a million young Americans at one time but made it possible for some major conservation projects to be undertaken and completed. For instance, it was reportedly responsible for planting three billion trees in the US over a ten-year period. Before setting it up, Roosevelt knew that urgent action was needed to deal with both spiralling unemployment and massive environmental degradation in the US. He got things moving with amazing speed by marshalling the heads of key government departments. As a result, the first CCC camp was established only weeks after Roosevelt signed a bill to create the program. Previously unemployed people joining the scheme were issued with ranger-style uniforms. By all accounts, a strong camaraderie developed between them. By contrast with military training, the CCC was seen as a peaceful yet positive program. It achieved tremendous results in the US in reversing deforestation and erosion.

I believe that with some small modifications the Australian Government could replicate the CCC. In my view it is an ideal blueprint for our time. It would give large-scale conservation work in Australia the impetus it desperately needs. In practical terms, it would also give the relatively small number of people who own and work our rural land—the nation's farmers—the help on the ground they need. In return, I am sure that those who joined the scheme, many of whom, probably, would have been unemployed before, would gain a great deal in knowledge and other ways from the farmers they work with. All concerned would benefit, the environment most of all.

25

HOW MANY TREES ARE ENOUGH?

Experts seem to agree on one basic requirement for farm sustainability: every farm should have a sizeable proportion of its land under trees and shrubs. The consensus seems to be that this proportion should be around 35 per cent. Several independent studies have come up with a figure in that order. Which leads to the next question: what species of trees should be planted? This is a matter I have been mulling over for the past three or four decades, and I have come to the conclusion that roughly half of the revegetation should consist of native (preferably indigenous) tree and shrub cover and the other half of trees grown for timber. The latter should include both native and exotic trees.

In 1999 Dr Sandra Walpole, an agricultural scientist, did a study in the Gunnedah area of northern New South Wales which examined how the productivity of a farm's pastures varied according to the proportion of the farm that was under woodland vegetation—that is, covered by trees, shrubs and associated ground flora, such as native grasses and herbs. In short, she set out to discover if trees made a farm more productive. In my opinion, every farmer in Australia should become acquainted with the results of Dr Walpole's research. She found that a farm's pastures reached a peak output when the proportion of the farm covered by trees and under-storey rose to 34 per cent. Any increase in tree and under-storey cover above 34 per cent produced no further increase in output.

This is how Dr Walpole summarised her findings:

> The results indicate that the value of pasture output per farm may be increased by having a certain proportion of pasture area under dry sclerophyll or woodland vegetation. Gross value of pasture output was at its highest level when the proportion of tree area across the farm was at thirty-four per cent with no further increases in output being achieved beyond this point. These results suggest that the competitive influences of trees present in the pasture system may begin to outweigh the beneficial effects when this proportion of tree area is exceeded. It is encouraging to observe that the thirty-four per cent value estimated in this study is within the range of values quoted in previous studies.

In other words, the optimum tree and under-storey cover on a farm should be about one third of the farm's acreage. Significantly, as she herself notes, other scientists have arrived at much the same conclusion. I know of four other papers based on research in the higher rainfall areas of southeastern Australia which conclude that the optimum tree and under-storey cover is in the 30 to 35 per cent range. If Dr Walpole and the other scientists are correct, all farmers should aim to have around a third of their land covered by trees and associated habitat. On ordinary family farms, though, achieving a tree cover of that dimension is a lot easier said than done, especially if the property is virtually devoid of remnant tree cover. This puts farmers at an enormous disadvantage—as I know from my own experience. If a farmer has remnant tree cover, he can simply fence off an area surrounding it and count on the existing trees to self-seed and revegetate the area naturally. If need be, the farmer can use herbicides judiciously to control competing grass. Without a remnant tree cover, farmers have to plant all the trees themselves.

This is what Cicely and I had to do at Lanark. And this is why, after 50 years of fairly intensive planting, I have still not got Lanark's tree and shrub cover up to 35 per cent. The cover is now approaching 20 per cent.

If farmers are hesitating about 'sacrificing' good grazing land for planting trees and scrub, the key thing to recognise is that, provided the right trees are planted in the right configuration, their farm's productive capacity will not suffer. Setting aside, say, 10 per cent of your grazing land for trees and scrub certainly does not mean you will lose 10 per cent of your farm's income. The trees and scrub you plant should allow you at least to maintain the same income by increasing the productivity of the remaining 90 per cent of your land in the manner I have previously described.

An important flow-on benefit is to make the farmer less reliant on artificial fertilizers like superphosphate, which represents a big financial saving. The farmer is also less likely to be reliant on herbicides, as has certainly been the case at Lanark.

Dr Rod Bird had a few interesting things to say about the pros and cons of planting trees on farms when interviewed by Warren Shaw, a landscape architecture student at RMIT. The interview was part of a research project centred on Vasey Farm, which then, as now, was run by my son Patrick. The interview took place 20 years ago, but Bird's views on whether farmers are better or worse off by planting trees are at least as relevant today as then. In his research report Shaw quoted Bird as saying:

> The farmer would put in, say, one or two kilometres of shelter belts per year. Then, depending on the total—say, if he had a 400-hectare property—he would end up with five to ten per cent of land completely fenced out to stock. Now, obviously if you are doing an economic analysis you've got costs and you've got benefits. An obvious cost is you're losing production from that area of land. The benefits are the direct effects of shelter on livestock.

Among those benefits, Dr Bird said, was a 5 per cent increase in the number of lambs weaned and a 10 per cent improvement in plant growth. For farmers, improvements like these can make the difference between a good and a bad year.

Only recently I saw one semi-government rural authority advising farmers to aim for a 5 per cent tree and shrub cover. Five per cent! To me, that seems a mere spit in the ocean. Having examined just about everything on the subject I have been able to lay my hands on, including Dr Walpole's research, and in the light of my own experience at Lanark, I am personally convinced that a tree and shrub cover of about 35 per cent, as a proportion of a farm's area, is the ideal. A farm that expands its tree and shrub cover to that figure will, on the one hand, have maximised its long-term sustainability and, on the other, gone a long way towards ensuring the health of the farm's environment. However, there is often a big difference between the ideal and the achievable.

Dr Bird argues that a tree and shrub cover of around 10 per cent is as much as any farmer can aim for in his or her lifetime, unless he or she is fortunate enough to have a lot of off-farm income or unless the Australian Government decides one day to help farmers meet the cost. In my view, though, we need to lift our sights environmentally: 35 per cent should remain our ultimate goal. Clearly, the damage done to the Australian landscape by land clearing over the past 200 years cannot be fixed quickly.

At Lanark, we are working to a 100-year plan that we adopted 35 years ago. If the property were to remain in the family, the plan would probably be seen to completion by my great-grandchildren. That may seem too far in the future to worry about, yet I believe strongly that all Australians, not only farmers, need to think in time frames of this magnitude if we are to achieve equilibrium in our fragile environment. This does not mean that subsequent owners would have to stick hard and fast to the plan. But if the plan is prepared with expertise (as Lanark's was by Rochelle Ruddock), its mapping and other documentation will remain an invaluable reference to the state of the farm at the time the plan was prepared.

The same applies to farm forestry—that is, planting trees that can eventually be harvested for timber as part of the overall farming operation. Farmers know that any timber trees they plant today will not be harvested until the 2030s or, depending on the type of tree, perhaps some years after that. Why lock up good farming land that you can use productively today in the expectation of cashing in when the trees are harvested at some far-distant date in the future?

It is a good question, for which there is an equally good answer. The point is that farm forestry does not require a farmer to plant out whole paddocks with trees. The trees a farmer grows for timber are best planted in strategically positioned blocks ranging in area from a half-acre to two acres (0.2 to 0.8 hectares). These blocks should be inter-connected with shelter belts, windbreaks, wildlife reserves and wetlands. If planted correctly, the trees will improve the farm's production while they are growing to maturity and, once mature, will provide an ongoing source of extra income. In short, a farmer who takes on farm forestry has nothing to lose in the short term (except the time and money it takes to plant and fence off the trees) and much to gain in the medium and long term.

I strongly believe that farm forestry should not be separated from the rest of the farm. Rather, it should be part of an integrated program of tree and shrub planting on the farm. The key word here is 'integrated'. Farm forestry trees should be only a component of a farm's tree and shrub cover. They should certainly not be grown as monoculture plantations—that is, in areas reserved solely for, say, radiata pines or blue gums.

Any farmer who intends combining habitat with a timber plantation in this way would be wise to allow the species of timber trees to determine the location of the plantation. For instance, if your choice of timber tree was river she-oak, you would look for an area with deep and rather damp soil, which river she-oaks like. Having identified such a site and planted the river she-oaks there, you would then add various habitat trees and shrubs that suited the same conditions. It goes without saying that the habitat plants should preferably be local—that is, indigenous—species. In my experience, most districts in Australia have at least one nursery that grows plants from local seed, and with luck you ought to be able to obtain your habitat plants there. In some parts of Australia—and certainly in Victoria—the local Field Naturalists Club (FNC) would probably be able to put you in touch with such a nursery if you cannot find one yourself. You can also draw on FNC members' knowledge of local flora and fauna.

The plant you need most when you set out to revegetate a degraded area is the local pioneer wattle—that is, the most aggressive of the area's indigenous wattles. In my own district, this is the black wattle. After a bushfire here, black wattles will be the first plants to come up. Or if you were to lock your stock out of a paddock for any length of time, you would find black wattles springing up—assuming there were black wattles growing somewhere nearby. If I wanted to establish a habitat at Lanark and there were no black wattles growing nearby, I would buy them from a local nursery and plant them. Better still, I would collect the seed and sow it directly. The advantage of wattles is that they grow rapidly and so provide shelter for slower-growing but longer-living species; they charge the soil with nitrogen; and, being relatively short-lived, they break down and hasten the development of a balanced environment.

Now, I know some farmers do not like planting black wattles because they have a relatively short life span. (It is not as short as you might think, however: at Lanark we have known black wattles to live for up to 30 years.) Still, there is no better plant for kick-starting the process of habitat reconstruction. The fact that they are not long-lived is actually a plus. It means you will soon have a supply of dead and rotting wattles—and dead and rotting trees are a vital component of any bushland reserve. Remember that when you set out to create a wildlife habitat you are not doing it only for birds and mammals. You are doing it for all native life; bugs, grubs and fungi included.

Remember, the area you fence off as a reserve should not be too small. I regard a half-acre as the minimum: an acre or more is preferable. Ideally, any area a farmer fences off for habitat should not have a gate. Why? Because if there is a gate, the farmer might not be able to resist the temptation to let stock in to graze whenever feed is scarce.

26

ALLEY FARMING

Alley farming is a method of farming between rows of trees that offers important benefits for both the farmer and the environment. It is a term that can mean different things to different people, yet the basic concept is simple enough. Some years ago a United Nations Food and Agriculture Organisation (FAO) report called alley farming 'an agroforestry system in which food or forage crops are grown in the "alleys" between hedgerows of trees or shrubs', which really explains it all. A recent CSIRO report offered a similar definition. It described alley farming as 'an agroforestry system that comprises belts of trees and/or shrubs, alternated across the landscape with strips of crops or pasture'. Alley farming has not yet caught on in Australia, but I believe that it will eventually. There is just too much evidence in favour of it to be ignored forever. One problem is that alley farming cannot be adopted as a farming method overnight. A farmer has to wait five years or more for the timber belts that form the alleys to grow to a decent height. The timber belts are, however, worth waiting for.

Alley farming has numerous benefits. It allows higher cropping yields, reduces soil erosion, protects stock from cold winds and provides habitat for wildlife. The shelter belt can eventually be harvested for timber.

A lot of research work has been done on shelter belts (some people prefer to call them windbreaks) to establish how effective they are in protecting grazing and cropping country from wind. According to a formula that seems to be widely accepted, land behind a shelter belt will have optimum protection to a distance equalling five times the height of the timber belt. So if you had a shelter belt with trees ten metres high, a 50-metre strip of land behind them would be well protected from wind. This formula is a useful one, for it helps a farmer to determine how wide his alleys need to be. Tree species, soil type and rainfall can also be taken into account. At Lanark, the alleys are generally 40 metres wide, which is narrower than the norm. We opted for alleys of this width because we were looking for quick results. Generally, alleys are made about 60 metres wide. This minimises the cost of establishing them, since, if alleys are wider, fewer will be needed in a given area. It also means that tractors and other machinery can be manoeuvred within an alley more easily.

Our involvement in alley farming at Lanark (or at least a form of alley farming) began in the late 1980s. Two government experts who happened to be friends of mine—Rod Bird and Arthur Lyons, a senior government forester—initiated the project. They offered to set up an alley-farming system at Lanark—we preferred to call it a timber-belt system—as a practical trial. It was these two men, incidentally, who first acquainted me with the formula for determining the width of an alley. After a couple of years of planning, including excellent design work by Jim Sinatra and his RMIT students, the planting got under way in October 1989. Six timber belts were planted in parallel formation in what was known as the Big Dam Paddock, thus creating five alleys. Each of the belts consisted of several rows of trees, including pines and eucalypts, plus indigenous shrubs. The shrubs filled the space below the branches of the trees, thus forming an effective windbreak. The shrubs offered another advantage: they provided extra habitat.

The longest of the timber belts ran for 400 metres and the shortest for about 250 metres. All had to be fenced off from stock. By any standards, it was a big project, and we would have struggled to complete it on our own, but fortunately the project had community support. An amazing array of individuals and organisations were generous enough to help. For instance, members of the Apex Club at Hamilton planted the trees under the direction of a soil conservationist, John Langford, a close friend of mine. A fencing contractor, Graham Bowen, aided by my son Johnnie, did the huge amount of fencing that was needed to protect the trees. It was a tremendous cooperative effort, and it would be wonderful if it could be repeated wherever there was a need. I could think of no better way of revegetating the landscape. The trouble is that this kind of community effort is no longer possible. Country people are innovative, practical and generous with their time, but they are hopelessly constrained today by health and safety laws emanating from the city, which may be well intentioned but which have the effect of stifling community activity.

In 1989, though, with so many people there to help, the project went ahead swiftly and efficiently. Its main purpose was to measure the benefits of having stock graze on pastures that were protected on either side by timber belts. We also wanted to find which silviculture techniques were right for our conditions and to see if the native hardwoods that we planted in the timber belts would prove valuable as a source of specialty timber. The project did not seek to assess the benefits of growing crops in alleys, although we did have this demonstrated to us at Lanark a few years later. I had agreed to let a share farmer (someone willing to plough and sow paddocks on our land in return for half the crop) to grow a crop of oats on a paddock of about 100 acres (40 hectares), which occupied a bare hill. As an experiment, I also asked him to crop the alleys. The share farmer was not too keen on the idea of working in the 40-metre-wide alley, saying he would find it hard to manoeuvre his tractor in an area as narrow as that, but as I insisted, he did it.

The results were remarkable—and nobody was more surprised than the share farmer himself. He harvested about a third more oats to the acre in the protected alley than in the adjoining paddock on the bare hill. The soil in the strip was virtually identical to the soil in the paddock, and both were prepared and seeded in the same way. Clearly, protection from the wind had allowed the crop in the alley to grow better.

Windbreaks are effective in preventing soil being blown away, a fact that ought to have enormous significance for more marginal cropping districts like the Mallee in Victoria, which have lighter soil structure and are prone to dust storms. Some farmers there have had a go at alley farming. They have to be able to manoeuvre their huge harvesters within the alleys, but, provided they make their alleys wide enough, this should not be a problem.

All of which leads to the question: if crops and pastures grow much better in alleys between windbreaks, why on earth are farmers throughout Australia not growing them there?

Sadly, the answer is simple: for most farmers, establishing the fenced-off timber belts would cost too much. Even on a medium-sized Australian wheat farm, the cropping area may well measure three or four kilometres from one side to another. If the farmer were to plant parallel windbreaks across an area of this size—even at intervals of, say, 100 metres—he or she would have to plant a colossal number of trees and, if stock were admitted to the area, erect a vast length of electric or other fencing. This would be an impossibly ambitious and costly project for most farmers to undertake.

Does this mean that we should simply give up on the idea of alley farming? I am afraid it does, at least in the absence of some kind of financial support for the farmer. I made the point before that alley farming is good for the environment as well as for crops and pastures. Imagine how healthier, more resilient and less salinity-prone the Australian landscape would be if multiple windbreaks became a standard feature of farming land. Imagine, too, how much more aesthetically pleasing our landscape would be, especially if the windbreaks were positioned in such a way that they were in harmony with the landscape. Clearly, the idea offers many important advantages, including the eventual sale of timber. In some cropping areas, admittedly, the trees will take longer to grow and the timber, when harvested, may not be top quality. Overall, though, the concept is a sound one, and, along with other sustainable practices such as minimum till, it is being adopted by at least a few innovative farmers.

My own belief is that many more farmers would be willing to give alley farming a go if the Australian Government provided them with low-interest loans to do it—or, better still, interest-free loans. As we have seen, assistance of this kind used to be available for farmers under the forestry loans scheme introduced in the mid-1960s. The term of the loans was fixed at 25 years, presumably because by then the timber plantings, assuming they were radiata pines, would be ready for harvest or close to it, thus providing the farmer with more than enough income to repay the loan. Of course, no bank would be willing to finance a farmer to establish a commercial tree plantation that did not produce a return for 25 years, which is another reason that government loans are essential. Indeed, to achieve what needs to be achieved I doubt if the Australian Government would need to outlay much more money in tax concessions than it did to support the blue gum industry.

In the meantime, alley farming must remain a pipe dream. Environmentalists can go on as much as they like about the desperate need for timber belts to prevent the wheat paddocks in Western Australia and elsewhere blowing away, or to alleviate the growing and frightening problem of salinity, but the stark reality is that the only people capable of implementing these measures, the nation's farmers, cannot possibly afford to put them into practice. The money is just not there. Most farmers would like to save the environment if they could, but their immediate concern, the one that keeps them awake at night, is how to earn enough money to hang onto their farms.

27

THE AMAZING VALUE OF SHELTER

As every farmer (but perhaps not every city person) knows, stock need to be protected from wind in winter because wind greatly aggravates the effect of cold. This connection between wind and temperature is often referred to as the wind-chill factor. We humans are well aware of the wind-chill factor. It is a matter of simple physics that in cold weather we lose heat more rapidly if our bodies are exposed to wind. Moreover, the stronger the wind, the faster we lose heat and the more we suffer from the cold. Sheep and cattle suffer from wind in precisely the same way. It has been found that even if sheep are insulated with a coat of wool three centimetres thick they will still be affected by wind. Hence the importance of windbreaks.

For newly born lambs, wind is a matter of life and death. As sheep farmers know well, lambs unlucky enough to be born on a day when it is cold, wet and windy are liable to die, especially if they are small or weak. Twins and triplets are particularly vulnerable. Some research has shown that lambing losses are halved if adequate shelter is available. At Lanark, the results have been even better. Lambing losses fell dramatically when our ewes began lambing in the sheltered paddocks. Today, thanks to the abundant shelter available, lambing losses have almost been eliminated, a fact that my son David confirmed for me just the other day.

My other son, Patrick, David's twin brother, has learned from experience about the importance of shelter. Patrick and his wife, Bronnie, run Vasey Farm, the property that belonged to Cicely's family, the Gaussens. I referred in an earlier chapter to a research report by an RMIT student, Warren Shaw. In this same report, Shaw recorded a conversation he had with Patrick about the trees at Vasey Farm. He asked Patrick why he thought trees were important on rural properties. Patrick replied, 'The most important thing in this area is the salinity . . . then shelter. Because of the wind storm of '87, a neighbour lost 2000 sheep in a matter of two to three hours.' How many sheep had Patrick lost during that same storm at Vasey Farm, a property with an abundance of trees and windbreaks? Patrick's answer: just 22 lambs.

Windbreaks are important on grazing properties for another reason: sheep and cattle eat more when they are cold. In fact, this is the way they keep warm: by stoking their body furnaces with more and more fuel. It follows from this that stock protected from wind will burn up less of their feed in trying to keep warm and, therefore, put on weight. Here is one statistic I have seen quoted: sheep will be 30 per cent heavier if the velocity of the winds that they are exposed to can be reduced on average by half. The bottom line is that farmers who have their paddocks sheltered from wind will need less pasture to get them through the winter in good health. This in turn has various implications for the farm. Farmers can run more stock, safe in the knowledge that there is enough feed for them. Or they can save on buying hay and other supplementary fodder.

Winter storms take the greatest toll on sheep. In my own district, there have been three storms in the past 30 years that caused mass deaths of sheep—in 1982, 1983 and 1987. Some farmers lost 1000 sheep or more in one storm, yet on neighbouring farms where sheep had good shelter, losses were minimal. In most cases, the shelter was provided by shelter belts of trees or woodlots. Occasionally, it was provided by remnant vegetation.

Lambing remains a big problem, however—at least for sheep farmers. Nowadays, most sheep farmers prefer late winter as their lambing season, although some still choose to lamb in the autumn. Farmers who do the latter face an obvious problem: providing the ewes with enough feed to get them and their lambs through the long winter months ahead when pastures have minimal growth. Lambing in late winter is not free of problems, either. On the contrary: over the years, certainly in many areas of Victoria, there have been horrendous losses of lambs at this time because of the cold.

Lambing losses are a very old problem, of course. As I remember myself, it was the subject of endless discussion in the mid-1960s and various government authorities became involved in the search for a solution. There was a lot of interest then in the idea of building special lambing sheds, rather like maternity hospitals, where the ewes could lamb safely, regardless of the weather. About this time I attended a field day in Hamilton organised by the Department of Agriculture, at which the subject of lambing sheds came up. I was a fairly young farmer then, but I stood up and said I could see no logic in spending a lot of money on lambing sheds plus the extra labour they involved when ewes could be protected just as well from the weather by a shelter belt of trees, which cost relatively little to establish. I went on to say that four years earlier my small daughter, Amanda, had helped me plant a belt of she-oaks, which were now providing really good shelter. The audience, consisting largely of farmers and department people, was unimpressed. They let me know that in their opinion growing trees to protect stock was impractical.

Which trees make the best shelter belts? Personally, I think drooping she-oaks are hard to beat as shelter-belt trees. They also happen to be indigenous to the Hamilton area, so I admit to being biased in their favour. When someone asked me a while ago which type of tree was my favourite, I replied it was probably the river red gum. But I do have a soft spot for drooping she-oaks, which I consider the signature tree of Lanark. Drooping she-oaks were widespread at the time of white settlement, but most were later wiped out. Early settlers cut them down and fed them to their stock when grass was in short supply. Or people cut them down to fuel bakers ovens. Even when the she-oaks regenerated later, they were eaten by rabbits. When I planted a dozen or so on our roadside in the early 1960s I planted each of them in the middle of a wild briar rose bush, whose thorns protected the tree from rabbits and hares.

The drooping she-oak is a superb tree in a number of respects. It is an evergreen, of course, and at Lanark, when fully grown, stands around six metres tall. It is a beautiful, weeping, fast-growing tree. Male trees bear large numbers of small, attractive, orange-brown spikes at the end of the branchlets, while female trees are festooned with large numbers of seed capsules or nuts. I used to call these the 'Lanark prickly men' when our grandchildren were small.

The reason she-oaks make such good shelter-belt trees is that they block just enough wind—and let just enough wind through. The latter is as important as the former. Windbreaks must never be impermeable. If instead of a shelter belt of trees you erected, say, a solid, ten-metre-high steel fence to protect your stock, you would generate so much turbulence on the leeward side of the fence that it would be almost as bad for the stock as the wind.

My policy when I began planting she-oaks as shelter for sheep was to plant them as if I were planting an orchard—in rows a few metres apart. I cut a hole in the middle of a superphosphate bag, lay the bag on the ground and planted a tree in the hole, the idea being that the super bag would keep the ground cool and moist, rather like a layer of mulch. I then put a tree guard around each plant, which I made myself from netting, and rotary-hoed around the plants for the first two years to ensure that no other plants were competing with the trees.

The reason I rotary-hoed was that there were no safe sprays on the market then to control weeds. The rotary-hoeing proved very effective. My aim was to keep the ground around each tree completely bare to a radius of 75 centimetres. When safer sprays became available, I used them instead of the rotary hoe. By now, we were planting trees not by the hundreds but by the thousands, so spraying was the only practical option. In areas where there is a dense growth of grass or weeds, the site of each tree should ideally be sprayed in the previous spring—a year before planting—to a radius of 75 centimetres. Do not blanket-spray. It is enough to have the ground around each tree bare to a distance of 75 centimetres. The rest should be left to provide some ground-level shelter and a microclimate of grass and weeds.

If you are not well enough organised to spray in the previous spring, then spray in the following autumn (you must wait until autumn to catch the weeds that germinate then) and spray again immediately before planting in spring. As I explained in an earlier chapter, we used to plant in autumn, as many people in some areas are obliged to do, but we found that the trees we planted then often fared badly. They would just sit in their holes for the whole of the winter, suffering from the frosts and looking sorry for themselves. This is why we switched to planting in spring, preferably late September or early October, when the ground was warming up. Provided we did the required preparation work, the spring plantings proved successful.

The real key to growing trees successfully is to do a follow-up spray the following autumn—that is, about six months after planting. Not many people do this, which is why not many people have trees that grow rapidly. Ideally, you should strive to keep the area immediately around each tree bare to a distance of 75 centimetres for the first eighteen months. If you do this, you should not need to water the tree after you plant it. Whenever I plant trees I always have the tube stock sitting in a bucket of water. If the ground is dry, I may also water each hole just before I put the tree in. After that, though, watering will not be necessary—provided the trees you plant are of a suitable species and provided they do not have to compete for moisture with weeds and grass.

Planted and fostered in this way, the she-oaks we planted at Lanark had close to a 100 per cent survival rate. They grew rapidly, too, generally reaching a height of around 1.5 metres in twelve months. The rapid growth meant that, if the she-oaks were fairly close together (at Lanark we planted them 3.5 metres apart), they would provide sheep with perfectly good shelter in as little as three years.

Drooping she-oaks have another attribute that appeals to me personally: the sound they make when wind blows through their branches. It is a whispering noise, although this description hardly does the sound justice; you have to hear it yourself to appreciate it. The nineteenth-century commentator James Bonwick was obviously impressed by the sound. In his book *Western Victoria, its Geography, Geology, and Social Condition: The narrative of an educational tour in 1857*, he wrote, 'Gentle undulating plains spread out before me, clothed with the weird-looking Casuarina or She Oak, through whose knotted, leafless appendages the wind breathes with a soft music like that of the Aeolian harp.'

Drooping she-oaks produce excellent timber. The wood is extremely hard and difficult to work, but it splits readily and was once used extensively for making shingles. The shearers' quarters at Bassett has a roof clad with casuarina shingles that is still intact. The she-oak's botanical name, by the way, used to be *Casuarina stricta*, but a few years ago this was changed to *Allocasuarina verticillata*.

The mention of the she-oak's botanical name reminds me of something that happened at a farmers' meeting at Branxholme some years ago. The guest speaker was a botanist, and during the meeting an elderly farmer got up and said he had a question to ask about establishing stands of *Casuarina stricta* in certain parts of his property. Instead of just answering the question, the botanist began by pointing out that *Casuarina stricta* was an obsolete name. The old farmer was made to look foolish. I thought this was intellectual snobbery of the worst kind, and I told the botanist so afterwards. I said, 'That's the last time that poor old bastard will ever try to use a botanical name.'

A friend, Dr John French, a CSIRO research scientist, urged me for years to establish a 'coppice-with-standards' plantation at Lanark in the traditional European format. He believed plantations of this kind would be a valuable addition to Australian farms generally. A coppice-with-standards plantation typically consists of timber trees (the 'standards') that are allowed to grow to full height in scattered formation within a fairly dense woodlot of smaller trees that are coppiced regularly—that is, heavily cut back for firewood or timber. In the spring of 1985, not long after John had visited Lanark, I began to plant just such a woodlot. When established, there were about 50 'standard' trees, consisting of various eucalypts, growing among 1100 drooping she-oaks, which formed the coppice understorey. We also included a handful of habitat trees, specifically river she-oaks and swamp paperbarks (*Melaleuca ericifolia*) in a particularly wet area. The entire woodlot occupied an area of not quite three acres. At the time, I could picture myself at some distant date sitting in the copse and listening to soft breezes stir the drooping she-oak branchlets.

In 1985 it was estimated that the standard trees would be harvested in about 60 years. In fact, quite a few of the standards did not survive. Instead, the best-formed of the she-oaks were pruned and allowed to grow to full height.

Windbreaks have one other role to play in the Australian landscape that most farmers seem unaware of: they can retard bushfires. How do they do this? Simply by reducing wind strength. An effective windbreak of trees will greatly slow the spread of fire downwind of it. I say 'greatly' because a reduction in wind strength always slows a fire exponentially. By that I mean that if a wind fanning a fire has its strength reduced by, say, 50 per cent, the speed of the fire's spread will be reduced by more than 50 per cent and probably by as much as two-thirds.

28

TREES LOOK GOOD, TOO

I would like to think that most readers, having got this far into the book, would accept that there are five sound reasons for planting trees on farms. One, they provide shelter and shade for stock; two, they provide shelter for pastures and crops; three, if the trees can eventually be harvested as timber they will provide the farmer with a future source of income; four, they provide a habitat for birds, insects and various other animals; and, five, they generate fertility.

The last of these is more important than I think most people realise. I have long observed—without any specific research data to confirm the observation—that trees enrich the soil around them. The obvious explanation for this is that the trees are bringing up minerals and other nutrients from the deeper soil and that these minerals and nutrients are being recycled in the soil when leaves fall on the ground. It is also a fact that trees trap airborne nutrient particles, including particles held in suspension in mists, and these particles eventually make their way to the ground beneath the tree and re-enter the soil there. The bottom line is that provided you had enough trees on your farm for long enough, your farm would be more fertile and more productive.

Personally, I believe there is also a sixth reason that is just as important as the others: trees are good for the farm aesthetically; they improve the look of the place. I have said elsewhere that I do not consider the aesthetics of tree planting to be a minor matter. I certainly do not consider it to be mere window-dressing. On the contrary, it is of immense importance. If you like the look of where you live you will be happier living and working there. This applies to farmers as much as it does to those living in the city. More so, probably. Visually attractive landscapes can also attract visitors from the city, so there can be a spin-off in terms of tourist dollars. City people want the countryside to be pleasing to the eye, and perhaps the day is coming when they will start pressuring government authorities to ensure that the countryside does please the eye.

My own policy is to plant trees in such a way that the landscape ends up looking as much as possible like an Australian landscape. Now, many farmers in our corner of Victoria have planted Monterey cypresses in single rows or double rows along the fence lines. In my early days at Lanark, I did some planting of this kind myself. If you were to fly over the settled areas of Victoria's southwest today, you would not think it looked like Australia. If anything, it looks rather like the Canterbury Plains when I first went to New Zealand in the 1950s, for farmers there had done a lot of single-row and double-row planting of cypress and pines along fence lines.

In the Australian landscape, trees typically grow at random in broad groupings, and in my view a farmer planting trees should make some attempt to simulate this. At the very least, farmers should have more than two rows of trees in each line or shelter belt. I prefer a minimum of five rows. There is a simple equation to consider here. If you plant five rows rather than two, you will more than double the number of trees you plant and thus more than double the benefit you get from those trees. But your fencing costs, far from doubling, will rise by a trifling amount. After all, the cost of fencing along the length of the shelter belt will be the same. It's only the width of the shelter belt that changes—from, say, eight metres to sixteen. So for an extra sixteen metres of fencing (eight at each end) you more than double the number of trees. In light of this, I have never been able to understand why anyone would plant only one or two rows in a line of trees when five rows would cost little more. The extra trees will ultimately be worth far more than the bit of land that they are planted on.

I have already spoken of Jim Sinatra several times in this book. Let me describe how we came to meet. I was at a football match in Melbourne in 1981 when my friend Graeme Gunn, then dean of the faculty of Architecture and Design at RMIT in Melbourne, introduced me to an American academic, Sinatra, whom he had just brought to Australia to work in the faculty. Sinatra's job was to headstart a new landscape architecture course at RMIT. He was eager to do a trip out of Melbourne as soon as possible to have a look at the Australian countryside, so I invited him to visit us at Lanark and not long after that he spent a few days with us. It was the first of many visits that Jim has made to Lanark over the years; often he has brought groups of students with him.

At the University of Pennsylvania Jim had studied for his masters degree under Ian McHarg, the renowned Scottish-born landscape architect, who was one of the first to promote the idea that the environment should be a fundamental concern in all planning. McHarg died in 2001 at the age of 80. About ten years before that he visited Australia and made a trip to Hamilton, where I met him at dinner at the home of my cousin, Richard Walter. He, too, had studied under McHarg at Philadelphia. It was quite an experience for me, because I had long been an admirer of McHarg and his theories, especially the notion that planning should conform to nature. I bought his famous book, *Design with Nature*, when it first came out in 1969 and was strongly influenced by it.

So it was not surprising that Jim Sinatra and I should share many of the same views on farm design. Jim had taught for fifteen years at a university in Iowa, traditionally a rural state, so he was naturally attuned to the special requirements of broad-acre design. In 1983 he set up a new program to bring his students into close contact both with rural Australia and with far-flung Aboriginal communities. As part of this program, called RMIT Outreach Australia, Jim would bring groups of students to Lanark at least once a year to inspect the tree-planting projects I had undertaken. The students who did these field studies were drawn from various years of the RMIT course, but they did have one thing in common: a special interest in broad-acre design.

Two of the students went on to produce brilliantly researched papers on Lanark and its environment. One of them was Elizabeth Jacka, whose study of bird life at Lanark I referred to in an earlier chapter. Elizabeth focused on how bird life had been affected by European settlement over the past 160 years or so. She looked at how bird numbers would have fallen when the land was cleared and drained, and native woodland were replaced with treeless paddocks, and how they rose after we began large-scale tree planting, the restoration of the wetlands and the creation of habitat reserves. The other student, Rochelle Ruddock, wrote and published a superb, detailed study in book form of the farm forestry operation at Lanark. For both students these were final-year projects, and both received honours for them.

Rochelle's paper (it was a project for her honours degree in landscape architecture) included a 100-year plan for Lanark—that is, a blueprint of how the property could be developed over the coming century. Now, some people may consider landscape design to be of little practical importance in urban areas. In the country, it is very important indeed. If a landscape design in the country has shortcomings, these will be magnified many times when the design becomes a reality. Moreover, whereas it may be relatively easy to correct a mistake in an urban landscape design, the scale of any mistake in the country is likely to be much too big to allow correction. Landscape design came into being as a separate discipline in the United States, and I was not at all surprised to learn that its original purpose was to improve the design of the rural landscape—that is, of farms—and to alleviate the soil and water problems that were starting to appear on rural land.

All in all, my association with RMIT University has been an enormously happy and productive one for me. It was particularly gratifying, therefore, when RMIT awarded me an honorary doctorate in landscape architecture in 1997. At the time I had no idea, but I now know that Dr Yaso Nadarajah, Senior Research Fellow at RMIT Globalism Research Centre, was the driving force behind this award.

Yaso had become closely connected with the work at Lanark, often visiting us with her group of research students. She is a wonderful organiser and managed to keep her plan a secret from me over the twelve months that the process took. Yaso approached Jim Sinatra, and together with Jim's wife Curtis, a good friend of Cicely's and also an excellent organiser, they became the triumvirate who hatched the plot. I had no idea this was in the pipeline and was totally shocked when I received a phone call from David Knight, secretary to the University Chancellery, asking me whether I would be happy to accept the honorary doctorate. No honour has pleased me more.

I have long believed that the skills of landscape architects have been underutilised in the country. I argued this as far back as 1984 when I was invited by the Australian Institute of Landscape Architects to speak at the institute's annual conference, which was held that year on the Sunshine Coast in Queensland. I submitted in my paper that, by and large, landscape design and planning principles had not been applied to the vast areas of rural Australia, where, as I argued, they were badly needed. Reporting on the conference, the journal *Landscape Australia* was kind enough to say that my paper was 'perhaps one of the most enjoyable yet thought-provoking' papers presented, and it quoted me saying the following about the problems of rural land degradation: 'The people who can help alleviate these problems—that is, landscape architects—are almost entirely living in urban areas and are therefore putting their efforts into resolving problems associated with an infinitesimal area of the Australian land mass.' At that time, as I also pointed out during my talk, rural land—that is, land used for grazing or agriculture—amounted to 67 per cent of Australia's land mass, yet across this vast area landscape architects had scarcely had any influence on design and planning.

Having said all this, let me make it clear that I am not suggesting for a moment that farmers need to hire a landscape architect to advise them on how to plan their farms. Farmers ought to be able to do the design work themselves, provided they have a reasonable sense of what sits well in their particular landscape, which basically means a good sense of proportion. Clearly, many farmers do not take design into account at all when they plant trees. Drive for ten minutes anywhere in the country and you are likely to come upon at least one planting eyesore—perhaps a couple of rows of inappropriate tree species running through a gully along a fence line, that are in conflict, visually, with everything else and appear to offer no environmental benefit.

Initial design work can be best done with the aid of an aerial photograph, one reason being that it allows you to view the place as a whole. Over the years I have used aerial photos of Lanark for this purpose. In fact, I still have some that were taken in 1944. It is not hard to obtain these aerial photos, by the way. The nearest office of the state lands department, or whatever the relevant department happens to be called, ought to be able to provide them. When you get the photo, throw a sheet of tracing paper over it and start playing around with ideas. A good design will emerge eventually through trial and error.

According to Jim Sinatra, designing a farm plan really amounts to fitting together all the various pieces. He wrote to me recently:

> Doing a farm landscape is like putting a jigsaw puzzle together. Each piece has a corresponding fit. That is how a landscape should work in rural scenarios: the puzzle gets put together over time. But that means that there is a clear image that we all try to perfect. McHarg's layered ecological approach to planning was tailored to see how the puzzle can fit together and is useful to planning farm properties. Farmers know their soils, wind, water

and livestock movements: pieces of the puzzle. Together the puzzle can fit together including ecological and, most importantly, economic positive outcomes that are necessary for good farm management, otherwise you can't afford your land improvements.

29

GOING IT ALONE

In November 1980 a conference was held at Melbourne University titled Focus on Farm Trees, which I regarded then, and still regard, as a watershed event. So far as I know, it was the first major conference held in Australia on a subject close to my heart: planting trees on farms to benefit the environment. Unfortunately, the organisers (few of whom, apparently, were farmers themselves) decided to stage the conference in the middle of November when most farmers, at least in southeastern Australia, were flat out shearing and cropping. As a result, there were not many farmers there. I did attend, however, and on the second night of the conference I went for a drink with a few others to a pub across the road from the university. One of the others, Charles Massy, was a young farmer from the Monaro district in New South Wales. I had not heard of him until that evening, but I was to hear a great deal about him later on, for he came to be regarded as one of the world's leading authorities on merino sheep. Indeed, his book *The Australian Merino*, a massive volume that runs to almost 1300 pages, is considered a classic—the definitive work on the subject.

I got talking with Massy in the pub and told him about my tree-planting program at Lanark and the various other projects I had undertaken there to try to restore the environment. He listened attentively, which I found encouraging, since most farmers I tried to discuss the matter with in those days had little or no interest in the environment and switched off as soon as I broached the subject. Then he said, 'Well, what you've just told me is quite remarkable, because it's almost exactly what I'm trying to do on my own place. Everyone tells me I'm mad. The stock agent tells me I'm mad. Even my father tells me I'm mad. So I'm very pleased to meet someone doing much the same thing who's obviously not mad.'

All this happened nearly 30 years ago, and I have neither seen nor spoken to Massy since that conference ended. I went back to Lanark and he went back to the Monaro district, and no doubt he battled away like me, each of us on our own, both trying to do what was best for the environment. What I missed—and what he must have missed, too—was the comfort of associating regularly with other farmers with similar ideas about how to manage the land. If we could have met as a group from time to time, we would have given each other's spirits a lift. We would have felt we were not alone. As it was, without some form of mutual encouragement, ours was a decidedly lonely row to hoe.

Farmers who are prepared to do things differently for the sake of the environment tend to feel they are on their own—that all other farmers and most government bureaucrats are against them. This is not true: there are farmers with similar ideas about sustainable farming all over Australia—farmers who are striving to achieve an ecological balance on their land and, at the same time, earn enough money to keep their farm going.

The trouble is, you never find them gathered in one place. They never have the comfort and reassurance of being part of a group. I knew of a few other farmers in the Western District who had a special concern for the environment and, like me, were trying to run their farms in a way that would be sustainable in the long term. But none of them lived within 50 kilometres of Lanark, so I would run into them only occasionally, maybe every couple of months. Often, we would meet at a sheep sale, but then the sheep would be uppermost in our minds, so we would not get down to exchanging views on how each of us was faring in our attempts to achieve long-term sustainability.

Take the case of farmers who grow trees for timber, which is certainly one way to achieve sustainability. In Australia, such farmers have little or no contact with each other. In New Zealand, by contrast, farmers who belong to the New Zealand Farm Forestry Association are a closely bonded group. They meet every month in the North and South Islands and once a year they all get together for a big national conference. They are able to do this because none of them has to travel very far. In Australia, distance beats us. Getting together is simply not practicable. The fact that young people are still leaving the bush and that, accordingly, the rural population is falling means that the isolation problem is unlikely to improve any time soon.

I used the term 'long-term sustainability' before. In fact, by definition sustainability needs to be long term. Just about any farmer is capable of running a sustainable operation in the short term, in the sense of keeping a lid on the farm's input bills. But what is the use of that? In an environment as fragile as ours—and there are few environments in the world as fragile as Australia's—it is essential that farmers do things in such a way that their land is at least as healthy 50 years from now as it is today.

Notwithstanding the tyranny of distance that tends to keep us apart, farmers with a deep interest in the environment do cross each other's paths from time to time. In 1999, for instance, I came into contact with a remarkable farming couple from Western Australia, Dale and Terri Lloyd. Their family farm, Eden Valley, is near Dumbleyung in the state's southwest. After Dale Lloyd took over the management of the property from his father, he and Terri set about restoring the land to good health by adopting biodynamic farming methods. Cicely and I have visited Eden Valley twice in the past four years, and we have seen for ourselves the extraordinary success the Lloyds have had in achieving an ecological balance. Environmentally, theirs is the best farm I know.

Let me quote what the Lloyds themselves had to say about their farm in a paper they prepared a few years ago, because I think it sums up the matter perfectly. They wrote:

> Biodynamic farming will not suit every farmer . . . income is forgone from not continually cropping farmland while grain prices are inflated and wool prices depressed, but the advantages are that the farm has never looked more alive, beautiful or stable. It is a pleasure to work and live at Eden Valley and all family members enjoy good health and abundant energy. For the Lloyd family the benefits far outweigh the costs, with all members wishing to continue farming without the use of chemicals and without further degrading the environment or losing more of Western Australia's unique flora and fauna.

30

SIXTY-FIVE WILD ACRES

In the mid-1960s, Cicely and I decided to establish a wildlife reserve at Lanark consisting entirely of native trees and understorey vegetation. We recognised at the outset that any reserve we created needed to be big enough to support sustainable populations of the wildlife we hoped to attract. In the end, we decided to set aside 65 (26 hectares) acres. By comparison—I say this for the benefit of readers who have trouble picturing just how big a 65-acre block is—Fitzroy Gardens in Melbourne are 62 acres (25 hectares) in area and the Royal Botanic Gardens in Sydney 72 acres (29 hectares).

The site we chose was all but bare. The lack of any remnant vegetation initially put us a a huge disadvantage. We decided to create the habitat in stages, starting with a nucleus of 30 acres (twelve hectares), which we named the Other Side. After fencing the area off, we planted it out with a variety of indigenous trees and shrubs, including black wattles, drooping she-oaks, blackwoods and manna gums, plus a suite of shrubs and ground cover.

Within the 30 acres was the old swimming hole that I had dug out for our children, known as the Other Side. Its attraction for Cicely and me was its distance from the house—about 400 metres. We would never have dug a swimming hole close to the house, of course, lest one of our small children wandered into it one day and got into trouble. As described previously, whenever a swim was called for, we would pack the children into the car and drive to the Other Side. In that way we conditioned the children to think of swimming as a special outing. In recent years the swimming hole has been generally dry, but as I write this the hole is again full of water, and David and his family are swimming there again.

That original 30-acre reserve has evolved over the years, and today it is an established bushland that is home to a range of native animals, birds, reptiles, frogs, mussels and fish. The animals include eastern swamp rats, eastern barred bandicoots and swamp wallabies. Bits were added periodically to the original reserve, and the area of fenced-off habitat is now around 65 acres. In fact, our son David fenced off the last addition only four years ago.

Today, the Other Side Reserve occupies about 4 per cent of Lanark's land, a significant proportion. Some farmers might consider it wasted land, but this is not the way we see it. Over the years, the Other Side provided a substitute for the holidays we could not afford. We not only swam there but had picnics and so-called treasure hunts. These were hunts for particular kinds of wildlife, maybe a pioneer skink or a blue wren, or even for some kind of rare fungi. The first to find the skink, wren or other designated 'treasure' would win the prize. It is a game our grandchildren now play there. Lately, blue-tongue lizards have been added to the list of 'treasures'. About seven years ago a volunteer wildlife carer named Pam Turner released five of the lizards in the Other Side Reserve. They were injured lizards that she had nursed back to health. The lizard colony appears to be surviving well.

Although the reserve is fenced off from stock, David does let sheep graze there at certain times of the year. The idea was suggested to us by Tim Barlow, an ecologist who visited Lanark some years ago and gave us excellent advice on managing the habitat. He made the point that for the habitat to function as naturally as possible we needed to replicate the grazing of the native grasses and herbs that would ordinarily be done by kangaroos and wallabies. So, depending on the conditions, David periodically allows weaned, drenched lambs to graze there for a short time in late summer.

Bushland reserves can also provide excellent shelter for stock if an emergency arises—as it did at Lanark in 1987. In late November that year we were hit suddenly by extremely cold winds from the east. This was remarkable in two respects. One, cold weather at the end of November is a rarity in our part of Victoria, which is why we always considered it safe to shear at that time. Two, killing winds in our district—strong winds that are cold enough to kill stock—invariably come from the southwest. Accordingly, we aligned our shelter belts to protect the stock from winds blowing from that direction. So when those freakish winds hit us in 1987 our sheep were largely unprotected. Moreover, a sheep's fleece naturally loses much of its insulating properties as summer approaches, so our sheep were all the more vulnerable.

I was still shearing that day when a stock agent called in and asked what I intended doing about the sheep in the front paddock. I replied there was no need to worry—the sheep there had plenty of shelter. 'You're wrong,' the agent said. 'I saw ten dead.' In fact, when we went to check we found the death toll had risen to about 30. Clearly, we had an emergency on our hands. My youngest son, Johnnie, happened to be home, and he asked, 'Father, what the hell are you going to do?' When I replied that I was unsure what to do, Johnnie said, 'Well, I know what I'm going to do.' With that, he took our sheepdogs plus a couple of RMIT students who happened to be visiting Lanark and rounded up the sheep. Then he cut a hole in the fence around the bushland reserve and, with some trouble, drove the sheep in. Once inside the densely vegetated reserve, the sheep survived. In fact, I think we lost only three more sheep that day.

In 1994 we were approached by the Melbourne Zoo and asked if we would be happy for the zoo to release a number of bandicoots in our reserve. These were no ordinary bandicoots. They were a species in serious danger of extinction—eastern barred bandicoots, so named because their hindquarters are 'barred' with light stripes. They were once common throughout the Western District. An old-time Branxholme resident, Arthur Price, once told me he used to see 'hundreds of the little devils', as he called them, when he coursed his greyhounds on a paddock at Bassett. But loss of habitat and local extermination by predators, mainly foxes and feral cats, had reduced their population throughout Victoria to mere hundreds, and it happened that most of the survivors were in the Hamilton district. The widespread use of potent chemical sprays in the 1960s also took a heavy toll.

I derived considerable satisfaction from the fact that our wildlife reserve, created virtually out of nothing, was one of only four sites in the whole of the state where zoo-bred eastern barred bandicoots were released into the wild. The others were wildlife reserves near Hexham and Colac, and a National Trust property near Skipton. By all accounts, Lanark was judged to be the best bandicoot habitat on privately owned land in Victoria. If nothing else, this demonstrated again that an individual farmer could make a big difference.

Did we make a difference as far as the eastern barred bandicoots were concerned? Well, the bandicoot population did grow for about eight years. Cicely and David could take much of the credit for that. They worked tirelessly with departmental officers to try to make the project a success. After that, though, the onset of drought caused the bandicoots' breeding rate to slow, since feed was scarce. Since then, sadly, they have disappeared altogether. Foxes seem to have been the main problem. The surrounding blue gum plantations harbour foxes and, in my opinion, have allowed them to proliferate, which has been bad news for small native animals like these. Today, the only surviving eastern barred bandicoots in Victoria are housed in fenced, fox-proof reserves.

Our experience with the bandicoots confirmed my belief that if a species is known to be endangered the number-one priority is to protect the existing habitat. I suspect that some academic researchers, as distinct from those in the field, do not understand this.

As I write, it is 50 years since we began planting trees and shrubs in our wildlife reserve—more than enough time, you might think, for the area to return to a state of ecological balance. In fact, we may have to wait another 80 or 90 years for this to happen. Earlier, I quoted my ecologist friend Neville Bonney as saying during a visit to Lanark in 1998 that the first trees I planted there might not start to senesce until they were 140 years old. It would take as long as that, he said, for the gum trees to develop the hollows in which so many birds and other wildlife like to nest. About one-fifth of Australia's native birds nest in tree hollows, a remarkable statistic, which puts the onus on farmers to retain fallen logs and old trees and stags. Nesting boxes are a poor substitute. It is the same with bats. I am sure most farmers would be astonished at the number and variety of bats that roost in hollow trees on their properties. Research suggests that a decline in bat numbers may be behind the growing problem of insect pests on farms.

Bonney, a strong advocate of planting native vegetation on farms, made one other important observation that day: he told me I had not planted enough scrub-type eucalypts such as swamp gum—trees that are generally regarded as inferior types of eucalypt but which have the advantage, as far as the environment is concerned, of becoming hollowed out relatively quickly, thus providing nesting places for wildlife in much less time than other trees.

About ten years ago I took a group of Aborigines from South Australia on a tour of Lanark. After walking around a few of the paddocks, we came to the Other Side Reserve. We entered the reserve by climbing through a fence, and as we did so one of the Aborigines, a young man, said to me, 'The land here is much softer underfoot.' Then he asked, 'Do you think this is how the land used to be when the old people were here?' He was referring, of course, to the original Aboriginal inhabitants. He was right—the ground in the reserve was definitely softer—although I had not noticed it before. It was nearly 40 years, after all, since hard-footed stock, sheep or cattle, had trodden on it regularly. During that time, fungi had re-established themselves in the soil, breaking down the tree litter and softening the earth. Even the eastern swamp rats had done their bit by burrowing a network of tunnels through the soil like giant earthworms. All this was, for me, further proof that the land was at last returning more or less to its natural state.

31

BLUE GUMS

As far back as 1971 the Australian Government decided that the environment was important enough to be placed under the control of a government department, the Department of the Environment, Aborigines and the Arts, as it was then called, now Environment, Water, Heritage and the Arts. Thus, for almost 40 years there has been a federal department supposedly overseeing the protection of the environment in Australia. It is an environment—the world's driest inhabited continent—that desperately needs protection, too. You would think, then, that in all this time government authorities would at least have put an end to the deforestation of Australia, which has surely been the biggest single cause of environmental degradation. But, no, the loss of trees goes on.

Clearly, back then the onus was on the Australian Government to do two things. One, dissuade farmers from clearing existing scrub and forests by offering them financial incentives to conserve whatever natural resources, especially trees, that were under their control. Two, encourage farmers, again with financial incentives, to plant trees on a large scale in an environmentally sensitive way. Successive Australian governments over the past 40 years have failed to do either of these things. What the government has done, by offering generous tax concessions, is to encourage investment in hardwood plantations, especially blue gum.

Have the blue gum plantations helped the environment? In my own opinion, they have been an environmental disaster. No surprise there: every monoculture is bad for the environment, and blue gum plantations are nothing if not monocultures. I further believe that blue gum plantations have had a disastrous effect on the social structure of the rural communities where the plantations have been spreading, all of which I find very sad. I will have more to say about this later.

My main objection to blue gum plantations, though, is that in many cases, perhaps most cases, they occupy what was previously first-class farming land. I am speaking here of land that had been pasture-improved over many decades—that had been seeded with sub clover and treated regularly with superphosphate, a finite resource that Australian farmers imported year after year at considerable cost. The combined effect of this, over time, was to build up the fertility of the soil to a point where the land had the potential to remain highly productive for years. Then, in the 1990s, the Australian Government offered generous tax concessions to people investing in blue gums, one of the tree species most preferred by the woodchip industry, and suddenly the blue gum plantation industry was up and running.

The reason I feel so strongly about the blue gum plantations is that there happens to be a concentration of them in the Hamilton district where I live. I know about them first-hand.

Why was my part of Victoria singled out for mass blue gum plantings? There appear to be three reasons: the rainfall is high enough and consistent enough; the soil is deep, rich and of a suitable type; and there is a deep-sea port nearby at Portland, where the wood chips can be shipped out.

Big business put a lot of money into the blue gum ventures. I have no doubt that the 'feel good' factor had a lot to do with this. Corporate people are sensitive to environmental concerns—they cannot afford not to be—and as well as being attracted by tax incentives they thought that by investing in blue gums they would be helping to green Australia, prevent salinity and create jobs in the bush. All in all, they would be seen to be doing their bit as good, environmentally responsible citizens. I agree the blue gum plantations have greened Australia, bits of it anyway. It is even possible they may have kept some salinity in check, although I question if many, or any, blue gum plantations have been sited in salinity-prone areas. But I do not accept that the plantations have resulted in a net gain of jobs. On the contrary: I believe the plantations have had a terrible impact on the social structure of some rural communities.

How have the blue gums done this? By replacing farms, where farmers and their families lived, with plantations, where nobody lives. Blue gum plantations are owned by corporate absentees. There are districts that have been virtually taken over by them, and where this has happened local communities have died or are dying; in some places there is nobody left even to man the local rural fire brigade.

Nobody realised this would happen at the outset. Then, it seemed likely that the plantations would be merely an adjunct to existing farm operations—that they would provide farmers with an extra source of income at a time when farmers needed every dollar they could get. What actually happened is that farmers sold their land, holus-bolus, to the blue gum companies—and the farms disappeared.

Dr Rod Bird spoke about this when interviewed by Dr Martin Mulligan of RMIT—an interview that is posted on an RMIT website. Asked if he had been a supporter of the blue gum industry initially, Dr Bird had this to say:

> Yes, I supported it because I saw it as one way of farmers diversifying their production. But I assumed farmers would lease their land for blue gum plantations rather than sell it, because that would be more profitable than running sheep, even if wool prices revived. We are talking here about annual rentals they would have received of $100 to $400 per hectare—depending on the rainfall and the distance from the port and the actual site conditions. So this was a good income with no work required from the farmer. And the blue gum could have been part of a broader strategy to diversify farm production. However, many farmers decided to sell their land to the blue gum companies, and they, of course, are only interested in planting blue gums. So this has not led to the diversification I had hoped for.

What saddens me about the way things have turned out is that it could all, so easily, have been different. As Rod Bird said, the blue gum companies could have leased the land they needed from farmers, which would have strengthened, rather than destroyed, rural communities. Or the Australian Government could have done something entirely different. Instead of offering corporates a big tax break to invest in blue gums, the government could have offered them the same tax incentive to invest in the rehabilitation of the Australian landscape. The corporates would have gone for it, too, because it would have had the same feel-good appeal. The money invested could have been distributed to farmers in the form of stewardship payments in return for work they did to restore the environment—planting trees for timber on a long-rotation basis, reinstating wetlands and creating wildlife habitats.

Blue gum plantations may well have been part of the scheme, too, although in a modified form. Instead of planting blue gums from fence to fence, the blue gum companies could have been required to plant them on a scale and in a pattern (preferably on the contours) that was sympathetic to the landscape. The end result: a healthier environment, a more attractive landscape, and a more robust and financially secure rural community.

Instead, wherever there are heavy concentrations of blue gums, we now have the exact opposite: a blighted environment, a dreary, featureless landscape and a rural community that appears in terminal decline. That is how I see things, anyway. Blue gums have a harvest age of only twelve or so years. So every twelve years the land where the blue gums are growing is stripped bare. Nothing is left; then they start again.

In recent years we have seen another change in land management in south-western Victoria: a growing number of New Zealand dairy farmers have taken up land in our district. The move makes obvious sense for them financially, since they can sell their land in New Zealand for close to double the price they pay for land here. By and large, the land they have been buying used to be dairy country but has for a long time been used for breeding prime lambs or fattening cattle. The New Zealanders have been coming over and snapping up the best of it.

One good thing that can be said about this influx of New Zealanders is that they have paid enough for their land to 'overcut' the blue gum companies, by which I mean they were willing to pay more than the blue gummers. So the farms they have bought have remained farms with farmers living on them. Moreover, in my experience the typical New Zealand farmer is efficient, innovative and hard-working—and nobody can ask more of a farmer than that.

32

NATIVES VERSUS DECIDUOUS

I have long been puzzled by the inflexible attitude some local government authorities and conservationists have to non-native trees. They seem to regard them as noxious plants that have to be eradicated wherever possible. Recently, this attitude manifested itself in my own town, Hamilton. Now, Hamilton has a creek running through it, called the Grange Burn, along which deciduous trees were planted by schoolchildren in memory of soldiers killed in World War I. Over time these matured, and many seedlings and suckers sprang up beneath them. The visual effect created by these trees, with their summer greens and autumn reds and golds, was always one of the town's attractions. The trees also provided cool shade in summer. Then, a few years ago, the local council and the district's Catchment Management Authority decided between them to poison and remove all the non-native trees along the creek. These were mainly elms (*Ulmus* spp), oaks, weeping willows (*Salix babylonica*), poplars (*Populus* spp), ash (*Fraxinus* spp) and pines. As far as I could tell from various inquiries I made, the primary reason for removing them was that the trees were 'exotic'—that is, they were not natives.

I could not follow the logic here. What was so terrible about non-native trees, I wondered, that they had to be exterminated? After all, trees like these were planted in Hamilton when the town was first established. They are linked with the town's early history. Moreover, most of Hamilton's non-native trees today are growing not along the creek but in private gardens, where, I am sure, they will continue to grow happily for as long as the town survives. So why go to the trouble and expense of removing the relatively few non-natives growing along the town section of the creek? In any case, few if any of the 'native' trees planted in Hamilton's parks and gardens are indigenous to the area, so it could be argued that they are exotic, too.

I went to the Catchment Management Authority's office in Hamilton to ask about it all. The officer I spoke to told me that exotic trees were weeds and that weeds had to be removed. 'Our policy is to remove all exotic trees in our catchment,' she said. 'They are all weeds and they will all be taken out as funding allows.' She explained that these 'weeds' reduced water flow in the creek, creating perfect conditions for blue-green algae blooms. I told her that, if this was the authority's policy, then the authority had gone completely mad. It should be obvious to anyone, I said, that the source of the blue-green algae and many other problems in Lake Hamilton was in the creek's upper catchment, where the creek was not protected and where sheep and cattle grazed in fertilized paddocks right to the creek's banks. 'That's the part of the creek you should be concentrating on,' I told her. 'Forget about the bit that runs through the town; that's harmless by comparison.'

I got nowhere. Later, I learned that a number of deciduous trees around a bridge would be spared. Why? So they would continue to provide a nice backdrop for wedding photos. In fact, as I write this, most of the trees in line for removal are still there. Happily, it seems the funds have not been available to allow the removal of the trees to go ahead. Long may the lack of funds continue.

Having said all this, I obviously have a strong preference for planting native trees on farms—ideally trees that are indigenous to the local area—over introduced trees, including deciduous ones. One reason is the relative ease with which they can be planted. Elsewhere in this book I describe the process of direct seeding, which makes it possible for one person to plant large numbers of native trees quite rapidly. Another reason is cost: native trees when bought in sizeable numbers are nearly always a lot cheaper than non-natives. A third, extremely important reason is that native trees provide a habitat for Australian wildlife, especially birds. A fourth reason is resilience: native trees naturally do better in our climate and environment than most introduced trees. For one thing, they need little if any watering.

The main reason I favour native trees, though, is that they really offer our only hope of revegetating the Australian landscape. It should be obvious to everyone that a successful restoration of the Australian landscape, which more than anything requires revegetation on a large scale, cannot be achieved by planting, say, hundreds of thousands of oaks or desert ashes. Only native trees can be planted successfully and in large enough numbers to do the job—and only native trees can create a landscape that looks Australian. I consider this last point to be extremely important.

Wattles are very effective at revegetation, both because of the speed with which they grow and their capacity to regenerate. As mentioned, I have a personal liking for black wattles if only because they are the pioneer wattles in my own area. Cicely, though, has a special fondness for Cootamundra wattles (*Acacia baileyana*), so we invariably included a few of them in each planting. About ten years ago, someone with an interest in trees was visiting us at Lanark when he happened to notice that we were growing Cootamundra wattles. 'What are you doing growing those things?' he asked, 'Surely you know they're an environmental weed in this district.' I replied that, yes, I knew they were regarded as weeds, but I had not found them to be a problem at Lanark. In any case, of all the trees we have planted at Lanark only one in one thousand have been Cootamundras.

My visitor remained unimpressed. 'It's still wrong to plant them,' he said. 'It's bad practice.' I replied that, whether or not that was true, the Cootamundra wattle was my wife's favourite tree and if I did not plant Cootamundra wattles she might leave me—or stop helping me plant other trees, as she once threatened to do—so I have kept planting them.

I began this chapter by arguing there was a place for exotic trees in Australian towns. What about Australian farms? Is there a place for them there, too? Certainly there is. For one thing, they have enormous visual appeal when suitably planted. Deciduous trees offer another advantage: they retard fires. They do this by absorbing radiant heat.

Finally, let me say a special word in praise of one exotic tree in particular—the oak. In the higher-rainfall areas of southern Australia, the oak is one of the best deciduous trees an Australian farmer can plant. It also happens to be a personal favourite of mine. Oaks are trees that Australians tend to associate with

Europe, especially England. You imagine them growing in cool, moist forests—a far cry from hot, dry Australia. The reality is that oaks grow well in some areas of Australia. Young oaks put down a deep tap root which is not unlike a eucalypt's. I have planted a large number of oaks at Lanark with close to complete success. Indeed, I have planted a small forest of oaks which is now around sixteen years old. I watered the oaks after planting, but that was all. There was no follow-up watering whatsoever, yet they all flourished.

It happens that one of Australia's leading experts on oaks is a near neighbour of mine, a twenty-minute drive away, near the village of Dunkeld, which lies almost at the base of the Grampians, the tail end of the Great Dividing Range. His name is Bill Funk, and I believe his family has been in the Dunkeld district since the mid-1800s. Bill and his family produce fine wool of high quality on around 1400 acres (560 hectares), which is more than enough to keep them busy, but over the years he has somehow found the time to collect trees from around the world, some of which are extremely rare. At last count, he had more than 200 species of oaks, easily the biggest collection of oaks in Australia. His contribution to our oak plantings at Lanark was immense. He was also the driving force behind the establishment of the Dunkeld Arboretum, which has become one of the district's chief tourist attractions. I view him as an ideal farmer: one whose expertise at farming is matched only by his love of trees.

Bill Funk speaks of trees 'occupying space', which I think is an excellent way to express a basic concept. He once came to Lanark to inspect a trial planting of trees that had been pruned and thinned by the scientist conducting the experiment. I said, 'They're not looking good, Bill.' He replied, 'Well, John, he's done the wrong thing. He's thinned them before they occupied the space, so you might as well start the trial again.' He meant by this that the trees had been thinned—and therefore opened to the wind—before they had really established themselves and occupied the space they were meant to occupy.

33

THE ART OF PLANTING FARM TREES

Most farmers intending to plant trees will run a deep rip—that is, deep enough to fracture the clay below the top soil—along the line where they want the trees to be planted. Farmers normally plant trees in the rip line—that is, along the line where the ground is broken—the idea being that the trees will find their feet quicker if their roots can spread through the broken soil. Moreover, water tends to get trapped in the rip lines, and having been trapped there it can easily penetrate to the bottom of the rip where the trees' roots are.

Experience has taught me to do things a little differently, however. As a general rule I do not plant right in the rip lines but, rather, slightly to the side. Why? Because I have found that rip lines in our heavier clay soils are liable to get either too wet from pooling water during winter or too dry during a dry spell, the reason being that water evaporates from them freely. I have also found that if you plant a tree just off the rip line its roots will bond better. Again, this is not a hard and fast rule. When it comes to planting trees, you cannot have hard and fast rules, for how exactly you plant a tree depends both on the site and on the species and size of the seedling.

For planting large numbers of trees in the form of tube stock I use a tool called a Hamilton Tree Planter, an implement I had a hand in developing. When I first started planting trees at Lanark I fashioned a crude tool that enabled me to open holes in the ground just big enough to receive the tube stock I was planting. It consisted simply of a length of rounded hardwood with a bolt through it, on which I could press with my foot to force the tool into the soil. Keith Cumming, then a local officer of the Pasture Research Institute, got to know of my planting tool, and he used it later as the basis for a more advanced piece of equipment of his own design, which came to be known as the Hamilton Tree Planter.

The Hamilton Tree Planter roughly resembles a garden fork, except that instead of the prongs there is a steel container the size and shape of an ordinary plant (forestry) tube. When you pull it out you remove a divot of soil the same size and shape as the tube. You simply drop the tree in the hole and tramp down the surrounding soil. As you drive the planter into the ground when planting the next tree, the soil already in the container will be forced out. I used this surplus soil to cover the top of the soil around the young plant. This surplus acts as a mulch, helping prevent the plant drying out by reducing evaporation from the potting mix in which the plant is growing.

We have used nurse trees in many of our forestry plantings. Nurse trees are planted to protect and shape the timber trees that you intend to grow. For instance, if you wanted to plant a group of she-oaks, you might alternate them with black wattles as nurse trees. The black wattles grow faster than the she-oaks, so they shelter them. If the black wattles are planted close enough, they will also force the she-oaks to grow up straight. The usual practice is to get rid of the nurse trees once they have done their job, although I do not do this myself. I usually leave the nurse trees, which, in any case, will probably die before long of their own accord, since they are invariably short-lived species.

Here is one example of a successful nurse-tree planting at Lanark as I recorded it in my tree journal. In the spring of 1984 we planted river she-oaks, a fine timber tree, along with black wattles as a nurse crop in what we named the Pulleyblank Plantation. In March 1994, I wrote in the journal that I had removed the biggest of the black wattles. So for almost ten years the she-oaks had been protected by the wattles and, as I noted in the journal, they were doing well. Obviously enthused by the success of the whole planting, I went on: 'We also have shelter, shade, landscape (aesthetics), firewood, untreated temporary fence material for on-farm pine agroforestry plantings, tops on ground for mammals, spiders etc., spars for the new chook house, shed for Murray Gunn's seed bank. All this from the black wattle nurse crop. The river oaks grow on—maybe for another 50 years.'

What you should never do when planting trees is put them out along the line at the required intervals before planting. Many people do this without realising that the trees' hair roots can get burnt off while the trees are lying there, waiting to be planted. Before planting, I would first stand the trees for a short while in a bucket of water and then carry them about in a damp jute bag.

On the question of watering: the fact is I did not water most of the tree species that I planted. I certainly did not water natives and conifers. I found that provided you supersaturated the tube stock before planting, by standing the tubes in water—and provided you had the correct planting technique—there was no need to water the trees once they were in the ground. One other proviso needs to be added here: you must ensure that the trees you are planting do not have to compete with weeds. This means that you must spray out the tree line before planting to kill any weeds, and you must maintain weed control for at least two years.

Of course, long before you start planting trees you have to decide where to plant them. A mistake farmers often make is to plant trees alongside an existing fence. In other words, they allow the location of a fence that may have been erected half a century ago to determine the location of the trees they intend to plant tomorrow. Now, this is fine if, by a nice coincidence, it happens that the best place to plant the trees is alongside an existing fence. After all, using an existing fence as a boundary for a row of trees does mean you will have one less fence to erect. But it is crazy to plant trees along an existing fence if this is not the best place to plant them.

I have been guilty of this myself. Jim Sinatra used to bring groups of his students to Lanark on field days. Before one such visit, I had pegged out the site of a row of trees I intended planting. I had it running down a hill alongside an existing fence. Jim Sinatra's students saw the folly of it immediately. Why plant the row of trees there, they asked, when it cannot possibly protect the sheep from wind, given that the prevailing winds blew from another direction?

Visitors to Lanark often comment on the fact that many of our smaller plantations of trees—areas of up to an acre or two (0.4 to 0.8 hectares)—have been fenced off in circular lots. Some are decagons with ten sides; others have 25 sides (icosapentagons). Why not simple squares and rectangles? It all has to do with the way sheep move around corners. If you plant your trees in a square or rectangle—that is, with right-angled corners—you will find that the sheep tend to wear out deep tracks around each corner which in time will lead to erosion. This is simply because the sheep repeatedly cut around each of the four corners while moving from one part of the paddock to another. But if you plant your trees in an icosapentagon, which is virtually a circle, the sheep tend to disperse, so the tracks they wear are not nearly so deep and damaging. Anyone wanting to know more about these circular configurations would do well to consult Rod Bird's 2000 book, *Farm Forestry in Southern Australia*.

There is another important consideration: appearance. Circular shapes look more natural and so fit into the landscape better. They certainly suit big, spreading trees like river red gums, for the fences forming the circles more or less approximate the trees' drip lines. Whereas many farmers plant red gums as stand-alone trees out in the paddock, we have always preferred to plant a number of them towards the middle of a decagon and surround them with wattles and various other habitat plants. Wattles fix nitrogen and are therefore good companion trees. Our hope is that 50 or even 100 years from now, when the fences forming the decagons finally fall down, a few red gums will be left standing.

I cannot stress this enough: the key to growing trees successfully is to eliminate competition. You must maintain good weed and grass control around the trees you have planted. Someone asked me recently if I used water crystals when planting trees. I replied that I considered water crystals a waste of money. Provided you maintain good weed control, water crystals are not necessary. Weeds are the number-one cause of young trees being deprived of water. I may add here that I do not mulch trees when I plant them, either. I used to do it. In our early years at Lanark, Cicely and I spent weeks mulching plants, but I realised eventually that it was not necessary. Again, provided your trees are of a species that suit the conditions and provided they do not have to compete for water with weeds and grass, they do not need to be mulched.

At Lanark, we made a practice of giving every planting a name. Usually, we named each planting after someone who had something to do with it. Even individual trees were given names. For instance, there is an oak on the place that we call Clive's Oak, the reason being that our oldest grandchild, Clive, planted it when he was a toddler. Naming plantings like this serves a practical purpose: it makes the plantings easy to identify and therefore easier to manage. It also personalises the plantings and reminds us of the friends and relatives who have helped us with projects.

34

THE MIGHTY RIVER RED GUM

Someone I know who lives not far from Sydney came to Hamilton recently with his wife and spent a few days touring the surrounding district. He had been here barely a day when he asked a question that I have heard from many other visitors over the years: 'What are those magnificent gum trees that I've seen growing around the place?' The trees he was referring to were, of course, river red gums, arguably the most famous of all Australian trees. Certainly, I know of no other tree that is so symbolic of the Australian bush, which no doubt explains why it has lately come to be described as 'iconic'.

River red gums do grow in the Hamilton area, especially to the north of the town, but the Western District could not really be described as typical river red gum territory.

Locals nearly always drop the 'river' part of the tree's name and call it simply 'red gum'.

As its full name implies, the river red gum likes to grow along and around inland waterways. It would be more accurate to say, though, that it is a tree that likes to grow where it has access to a shallow water table or where there is a good depth of subsoil capable of storing water. In inland areas, they grow in the floodplains that spread for some distance on either side of each waterway. This used to mean that the river red gums would get inundated once or twice a year when the floods came down. This rarely happens today—certainly not in the Murray–Darling river system—which is one of the reasons environmentalists are so concerned about the tree's survival.

I share this concern. I have no doubt their numbers will decline sharply over the next few decades if special steps are not taken to protect and nurture them and encourage natural regeneration. If that sounds like just another alarmist prediction, I can simply say it is based on my personal observation. The fact is that river red gums are already disappearing from the landscape at an alarming rate. I have seen this happen in my lifetime. How could it be otherwise when much of their ancient habitat has disappeared, too?

So what is so special about these trees? This is a question only someone who had never set eyes on a mature specimen would ask. Similarly, I find it hard to explain to anyone who has not seen a river red gum—and many city people have not—why the tree is so revered. The odd thing is that photos never do the red gum justice. You have to see the tree for yourself—standing massively in a paddock or river flat and dominating all that surrounds it. The red gum's size alone, especially the girth of its trunk, is awesome. They look old, too—and generally they are as old as they look. Some river red gums are known to be over 500 years old, and it is thought they may live to 1000 years or more. Bill Funk of Dunkeld, whom I have referred to elsewhere in this book, expressed all this well. 'The red gums are Australia's equivalent of oaks in

England,' he said. 'They typify the stout-hearted pioneers of this country.'

In 2009 a former Premier of New South Wales, Bob Carr, wrote eloquently about these mighty trees in the *Sydney Morning Herald*. He said:

> From the air they are bands of green that emerge from the Great Dividing Range and run along the banks of our major rivers as they snake west. On ground they are stately silhouettes, mighty arms akimbo, hefty trunks streaked red on white: statuesque, unmoving, some hundreds of years old but with the look of eternity. Hans Heysen's watercolours portrayed the river red gums as the guardians of inland Australia. He wrote: 'The red gums in the creek beds fill me with wonder; their feeling of strength of limb, of vigour and life, suggest the very spirit of endurance.'

Carr strongly opposes the logging of mature river red gums in the Riverina. I agree with him, although I do think we should plant them for future harvesting in areas where they occur naturally. The timber they produce is excellent in quality. The wood is hard, dense, colourful and durable. Over the years, river red gum timber has been put to a diversity of uses ranging from fine furniture (table and bench tops, for instance) to fence posts and house stumps. In Hamilton, I know of houses with river red gum stumps more than a century old. They have also been milled as railway sleepers. True, it may be many years before the trees are ready for harvesting, but that should not matter. Someone will harvest them eventually, and in the meantime the trees will be an adornment to the landscape.

Vasey Farm, the family property that my son Patrick runs, lies on the Dundas Tableland, just north of the Dundas Range—in the heart of red gum country. In fact, as mentioned elsewhere, there is a particularly heavy concentration of red gums on Patrick's property, and we can thank his grandfather, Paddy Gaussen, for that, for he did a great deal to promote their growth all those years ago. Even so, Patrick and his wife Bronnie have been doing their best to establish even more of the trees, and have had considerable success with a simple planting method. Patrick selects a few mature red gums, determines the direction of the prevailing winds and fences off maybe an acre or a half-acre (0.4 or 0.2 hectares) of land just downwind of the gums. Finally, he sprays out the fenced-off area with a herbicide to eliminate competition for the red gum seedlings that he hopes will sprout there. That is all he does: he does not plough, rip or in any other way try to work up the soil. He simply waits for seed from the red gums to be blown into the fenced-off area and to germinate there.

This could happen in months, or he may have to wait two or three years, for red gums tend to seed prolifically every second or third year. But it does happen sooner or later. Patrick's objective is to re-create more or less the kind of landscape that existed in the red gum country before white settlers arrived. By all accounts it was then open woodland—that is, one consisting basically of river red gums and grass. The Aborigines probably made sure by their routine burning of the landscape—fire-farming, as it is sometimes called—that the country did not become a jungle of river red gums. After all, it was obviously in their interest to maintain an open woodland where the game they depended on for food would come to graze.

The trees also act as windbreaks. Not long ago a wind abatement test was conducted in a paddock at Vasey Farm densely wooded with red gums. The result amazed us all. It was found that the wind there was only one-tenth as strong as the wind in an open paddock nearby.

Recently, Patrick prepared a report on Vasey Farm's river red gums—and their role in keeping salinity at bay. He wrote,

> When the ring-barkers came through 80 years or more ago my grandfather made them leave a lot more trees than some other properties. You look at the massive timber in one of those red gum trunks. Touching fingertip to fingertip you can get five people around the trunk. If all those trees started dying at once, this land would turn into a wasteland. There is no way that you could control the salting: there would just be groundwater coming out all over the place. One of those old red gum trees would be pumping about 200 litres of salty underground water per day, and up to 500 litres per day in the spring. It could take more than 1000 six-foot saplings to equal that.
>
> We lose about two big red gums a year from either wind or lightning. I watched a tree come down one day which probably had twenty tonnes or more of wood in it. Twenty tonnes of tree coming down. As you watch the tree fall in the wind you feel like getting a peg to hold the thing up, or to try and stop it halfway, because the tree is something so big and so beautiful.

River red gums seem to prefer growing in lighter country—that is, country that has sand in its soil profile—where they can get their roots down into moist subsoil. This is one reason they grow much better at Vasey Farm than they do, say, at

Lanark, where the soil is heavier. This is notwithstanding the fact that the average annual rainfall at Vasey Farm is around 100 millimetres less than at Lanark. It is drier at Vasey Farm and the soil is lighter, yet the river red gums at Vasey Farm are two or three times as big.

Is the river red gum my favourite eucalypt? Well, let me say I am also a great admirer of the wandoo (*Eucalyptus wandoo*) in Western Australia. I saw them fully grown for the first time when we visited that state in 2005. I remember coming upon a stand of them growing beside a roadside car park. For several minutes I stood transfixed, gazing at them through some light rain. Wandoos make a wonderful spectacle, and I thought at the time that whoever built the car park showed a lot of foresight in siting it there. In fact, by an odd coincidence, I had already planted wandoos some years before at Lanark, in 1987. My son Johnnie was one of the team that planted them. We mixed about 50 of them with a variety of other trees such as radiata pines, black wattles, drooping she-oaks and swamp paperbarks, but we named it the Wandoo Plantation.

The wandoos are truly superb trees. They are tall, often growing to 30 metres or more, and they have exceptionally attractive bark. There is a lot of white mixed with what I would call a Marmite colour. But, no, as much as I admire the wandoos I could not rate them above the river red gum. It is the grandest of all eucalypts. Red gums are living sculptures, really. They have another odd attraction for me. If you look around under a river red gum you will find wonderfully knotted and gnarled scraps of wood that have dropped from above. I sometimes take my grandchildren Lucy and Jack to an old red gum growing near the Vasey Farm homestead where they live, and they like to collect the interesting bits of wood they find beneath it. These come in extraordinary shapes; one bit will look like a racing car, another like a koala. If I ever took up whittling, which I would like to do one day, I would make sure I had access to a river red gum. For a whittler, there could be no better source of raw material.

35

LIFE AFTER LANARK

In the early 2000s we Fentons, as a family, had some big breaks with the past. In 2003 Cicely and I decided to leave Lanark and move into a house we owned in Hamilton. By then, David and his wife Catherine had been living at Lanark in a house of their own for close to ten years, and long before that David had taken over the running of Lanark's farming operations. As I mentioned earlier, he took over the farm's management as far back as 1992. So for a number of reasons it made sense for us to leave Lanark, our home for almost 50 years, and become town people.

All this while my sister, Robin, had continued to own a portion of Lanark. In 1998 I helped her organise the sale of her portion, part of which was bought by David. In 2000 Robin died, aged 61. More than twenty years earlier she had been given just three months to live. She used to say to me, 'You don't fight cancer: you just have to learn to roll with the punches.' She was always willing to try any new treatment that her excellent doctors recommended. She was also very supportive, in a quiet, reassuring way, of other women in the community who wanted to yarn with someone about how to handle an illness. Each evening Robin went for a long walk along the rugged cliffs at Cape Nelson, near Portland, which I think was an important outlet for her. She made a habit of stopping at a particular place along the walk and sitting there for some minutes in silent thought. For Robin, it would have been a time of meditation. After she died, her family erected a superb stone seat at the spot as a memorial. It contains her ashes and is adorned with a plaque.

One reason Cicely and I left Lanark, I do not mind admitting, is that the eastern swamp rats had made life almost intolerable for us, destroying our vegetable garden, ring-barking many of our oaks, even killing fruit trees. The odd thing was that about six months after we left Lanark the rats left, too. It was as if they decided the battle had been won—they had driven us away—so there was no point in fighting on.

The main reason for leaving Lanark, though, was that Cicely and I decided that with me looking over his shoulder David could not run the property as successfully as it could be run. This is a problem that arises quite often, I suspect, when farms are passed from father to son. It seems fathers cannot resist poking their noses into what should now be their son's business. The solution is for the father to move far enough away from the property that he cannot interfere on a daily basis. As I write this, it is six years since I left Lanark. I sometimes think that, for everyone's sake, I should have left ten years ago.

Our departure followed a disagreement I had with David over the way the property was being run. We had an argument one day, standing halfway between the house and the shearing shed, which ended with him telling me that the half-dozen or so major decisions affecting the property that I had made recently had been bad ones. Some colourful language was used—language that David had obviously picked up from me. I returned to the house to complain to Cicely. With her customary good sense, she came up immediately with a solution to the problem: she and I would move out of Lanark and let David and Cath run it on their own.

Leaving Lanark was undoubtedly the right thing to do, but it proved to be very difficult for me emotionally. Too much of my life had been invested there for the parting to be easy. For a while, I was quite disconsolate. To be frank, I was sometimes reduced to tears. I became particularly upset each time I went back to Lanark, for we still had our old home there containing a lot of our personal items, including most of my natural history library. In fact, if I had not been a farmer, I might well have concluded that I was suffering from a form of depression. But farmers do not suffer from depression—or so farmers like to imagine. In reality, depression has been rife among Australian farmers in recent years. Drought and financial hardship have combined to ensure that.

A friend of mine said to me just the other day, 'It must give you tremendous personal satisfaction to drive around Lanark and see what you have accomplished there. All those trees! And all that biodiversity!' I did not tell my friend this, but the truth is that five years after leaving Lanark I still cannot bring myself to drive around the property, for I would find that too hard to handle emotionally. My visits have been confined to our old home and its immediate vicinity. This will not last forever, though. The emotional stress of it all has been easing gradually. By the time this book is published, there is every chance I will have revisited Lanark in its entirety.

It is a matter of considerable pride for me that Lanark has now been passed to a third generation of the Fenton family. David is not engaged in any major tree-planting program, but in other ways he has carried on the work I began. He greets the visitors to Lanark who come periodically to look at and learn about trees and water, for instance, and I have been told that when he talks to the groups, he says the same things as I used to say and even sounds a little like me. That does not surprise me. After all, David grew up listening to me giving the same kind of talk to the same kind of groups. What I can say about David is that, as a farmer, he has his feet planted more firmly on the ground than I did. I was always thinking too much about the trees I had planted or was about to plant to give the farm the attention it deserved.

It took a long time to achieve, but when I left Lanark in 2003 it was functioning, in my opinion, as an environmentally sustainable property. Everything was in balance, by which I mean that the various elements of the environment—pastures, trees, sheep, wetland, wildlife, native scrub—were in equilibrium with each other.

One reason things were in balance was that we did not force the property to carry too many sheep. Today, David is running more stock than we did—not because he wants to but because he believes he has to if the property is to remain viable. On the whole, though, David is continuing to run Lanark along the same environmental lines as we did. Recently, I heard him argue the point with another farmer about weeds. The other farmer spent a lot of time spraying and slashing weeds, which meant his sheep ate little of anything but lush pasture grass. I heard David say to him, 'You wouldn't feed your children nothing but chocolate, but that's what you're doing with your sheep. You're feeding them rich food, without weeds to balance the diet.' I might have said that myself.

David believes that a combination of native pasture, improved pasture and weeds provides a balanced diet for his sheep, which consequently are generally healthier. He fertilizes only when soil tests show there is a specific need, and he always does it with trace elements and soil nutrient balance in mind. The basic aim is to maximise production while ensuring the health of the environment. David runs 3500 ewes, including self-replacing merinos, that are joined to Border Leicester rams and produce 500 first-cross ewes. The net result is the sale of 3200 prime lambs in late spring and early summer. Unlike me, David is an excellent shot with a rifle, and I remarked that he must have managed to shoot a lot of foxes, which are a principal cause of losses of young lambs. He said, 'Well, that's part of it. The main thing, though, is the shelter. We can lamb anywhere now without losing lambs.'

I came upon a similar quote by David in a book published in 1999, *Listen to the People, Listen to the Land*. He said, 'The trees have been a major part of Dad's life, so therefore it's had to influence us kids—and it has influenced us—I mean, you can see that. I don't think there will ever be salinity problems at Lanark because Dad faced up to the problem so many years ago. It's all about caring for the land, isn't it? That's why the trees are so important.'

When I look back now on all the work we did over half a century at Lanark to try to rehabilitate the landscape, I can see more clearly than ever before that it was Cicely who made it all possible. All the way through she was the anchor. She was the one who held things together. True, I did the bulk of the tree planting and other environmental work, but to a large extent I was merely indulging a personal passion. It was Cicely who had to bear the real burden of it all. She had to cope with the strain, both emotional and financial, that my passion for restoring the environment imposed on the whole family. I had the fun part. She had the drudgery and, worse, the worry. She also bore the burden of the bookwork. I always had an aversion to anything to do with accounts, budgets, bank statements. Cicely probably did not enjoy dealing with them, either, but she took them on for my sake. She was always the bookkeeper.

In fact, our achievements at Lanark were a genuine family effort in the sense that the burden was shared by our children, too. When we were awarded the 2002 Serventy Conservation Medal, named in honour of the great naturalist and scientist Vincent Serventy, I was pleased that the citation read 'John and Cicely Fenton and family'.

Neville Bonney wrote a paper about Lanark and its trees for inclusion in a book which, in the end, was not published. Neville was kind enough to give me a copy of the paper, however, and I would like to quote from the last page, where he describes a conversation we had a few years ago. While I was rabbiting on about the need to treat the land with a kind of spiritual reverence, as the Aborigines do, Neville broke in and said:

> John, look at the legacy you and Cicely have left, the many accolades you both have shared, the honours you have received during the past fifteen years. Lanark, the showcase for on-farm tree planting in the 1990s, has inspired many Australians to do the same. Lanark was the model farm. Now there are many Lanarks evolving throughout the landscape. Feel proud that Lanark was

> their inspiration . . . Lanark is now in its second innings, and it's still in the family, with more tree stories to come.

I hope he is right. Australia needs as many tree stories as can be told.

APPENDIX 1

Botanical names

Ash	*Fraxinus* spp
Black wattle	*Acacia mearnsii*
Blackwood	*Acacia melanoxylon*
Blue gum	*Eucalyptus globulus*
Cootamundra wattle	*Acacia baileyana*
Drooping she-oak	*Allocasuarina verticillata*
Elm	*Ulmus* spp
Manna gum	*Eucalyptus viminalis*
Mexican cypress	*Cupressus lusitanica*
Monterey cypress	*Cupressus macrocarpa*
Oak	*Quercus* spp
Poplar	*Populus* spp
Radiata pine	*Pinus radiata*
River red gum	*Eucalyptus camaldulensis*
River she-oak	*Casuarina cunninghamiana*
Sallow wattle	*Acacia longifolia*
Silver banksia	*Banksia marginata*
Silver birch	*Betula pendula*
Spotted gum	*Corymbia maculata*
Sugar gum	*Eucalyptus cladocalyx*
Swamp gum	*Eucalyptus ovata*
Swamp paperbark	*Melaleuca ericifolia*
Sweet bursaria	*Bursaria spinosa*
Tuart	*Eucalyptus gomphocephala*
Wandoo	*Eucalyptus wandoo*
Weeping willow	*Salix babylonica*
Willow	*Salix* spp
Woolly tea-tree	*Leptospermum lanigerum*

APPENDIX 2

Review of Lanark

By M.G. Cook, BAgSc(Melb) BEc(ANU)

The soils of Lanark are derived from basaltic rocks and are similar to those of much of the Western District. Most of the soils are deeply weathered, and can be described as loamy podsols in which the occurrence of ironstone gravel, referred to as 'buckshot', is a common feature.

The climate can be described as typical Mediterranean with cool wet winters and hot dry summers. The annual average rainfall for Hamilton is 27 inches (670 mm). The district of Branxholme would be expected to receive slightly higher rainfall. The timing of the autumn break is a critically important determinant of winter pasture production. Late summer or early autumn rains, if not followed up, can lead to some loss of annual species and if the first effective rainfall occurs in May then the pasture growth is seriously retarded by cool winter temperatures.

The breed of sheep carried were basically Corriedale, which is popular in the area. In times of high wool prices, Merino rams could be used to produce a comeback type, but generally the Corriedale was favoured because of the more saleable lamb or wether sheep when compared to the Merino breed. Crossbred sheep were introduced into the wetter and more fertile areas of the Western District when improved pastures replaced native grasses.

Pastures sown to improved species, notably subterranean clover and perennial rye grass, increased dramatically in the decade from 1930 to 1940, from 3 million to 16 million acres (1.2 to 6.5 million hectares) in the Western District. Of greater importance was the natural spread of clover by stock (via cattle droppings or by clover burr clinging to wool) or by feeding pasture hay. It was also necessary to topdress with superphosphate to overcome the extreme deficiency of phosphate in the soil. Perennial rye grass was sown from 1920 onward and was the grass of choice for much of the Western District. In the absence of perennial grasses, increasing soil fertility led to the intrusion of annual plant species such as capeweed and barley grass into pastures. This was undesirable because the presence of annual grass seed in the fleece devalued the wool, as did the dust contamination arising from areas left bare through the summer.

Lanark had a long history of topdressing with superphosphate, with the consequence that most areas would have received between 10 and 15 cwt of fertilizer per acre (1.27 to 1.88 kg per hectare) by 1964. Ryegrass comprised a dominant proportion of the pasture, although some annual weeds were evident, particularly on stock camps and areas unfavourable to ryegrass. Some phalaris had been sown under a cover crop, but later sowings were more successful where phalaris was sown alone.

During the 1960s, the universal consensus was that increased stocking rates were the key to profitable farming. This was embodied in the '4 Ss rule'—super, sub clover, subdivision and sheep. In the Survey of the Sheep Industry in the Western District for the year ending 31 March 1946 for the area centred on Hamilton, 62 per cent of properties had a stocking rate between 1 and 2 dry sheep equivalents (DSE) and 34 per cent carried more than 2. In 1964 the stocking rate of Lanark was just under 3 DSEs per acre.

During the period 1964 to 1970 it became obvious that property management had to become more sophisticated if farms were to meet the aspirations of the manager. In this regard the emphasis had to be placed on 'the 5 Ds'. These were: definition of goals, development of strategies to achieve these goals, and a risk management strategy to apply to drought, disease control and diversification of sources of income.

A major determinant of farm profitability from a grazing enterprise is the gross value of production. Sheep enterprises produce two commodities—wool and meat. The contribution each of these makes to farm income is a composite of quantity multiplied by average price per unit of output. With the emphasis on producing meat in the case of prime lamb enterprises, the key factors over which a manager has most control are numbers of lambs sold and their average sale price. For Corriedale or Comeback breeds the same holds true except that the age of sheep at sale may be older than lambs. By mating a portion of the ewes to British-bred rams, a supplementary prime lamb enterprise can be added without affecting the replacement of the primary ewe flock or the major characteristics of the wool clip.

Managing risk is essential for profitable farming. Some risks are definable in terms of the past experience of the farm manager. For example, the probability of a particular-sized drought occurring and the consequent fodder requirement for a particular stocking rate can be predicted. Other risks are less definable, but by addressing the question 'What if a particular event occurs?', sound strategies can be devised—both before and during such an event.

The 1967 drought was a watershed for properties in the Hamilton region. Most managers would have experienced periods of low rainfall—either a late autumn break or a lower than normal spring rainfall. What was not expected was to receive so little rainfall in both the autumn and spring of the one calendar year, the occurrence of which could be expected once in 50 years. The autumn rains did not eventuate, and even the month of July failed to bring effective rain to Hamilton for the first time on record. Although some rain was received in August and September the rainfall for September and October, which is critical for spring growth, was well below average. Another factor was that a failure of spring rains is much less probable than that of receiving a late autumn break, but when the autumn drought continued into spring there was no way of predicting when the 1968 autumn rains would occur.

From a business point of view the effect of the drought is to reduce the level of production. Selling stock may produce above normal income but the cost of feeding retained stock will increase (except at very low stocking rates). Regardless of the strategy pursued—that is, either selling stock or buying feed or both—the loss to the business is considerable and incurred both within the drought period and after the drought breaks. A manager faces a choice between increasing costs in the short term in order to return to full production when the drought breaks or avoiding the feed cost by selling stock in the drought but then purchasing or breeding stock after the drought. Some modelling of a drought strategy had been carried out for Lanark. It was not economic to maintain a fodder reserve sufficient to carry all stock through a drought of 1967 magnitude. Holding a fodder reserve represents a cost because the value represented by the stored hay or grain represents a lost opportunity of selling the reserve and making a profit. Depending on the assumptions used a rule of thumb was to hold a fodder reserve equal to half the maximum needed. In 1964 with the

Lanark property carrying 5229 DSE, the most economic fodder reserve would have been 3 bales of hay per DSE.

For any livestock enterprise, the risk of a disease outbreak must be considered. The 1960s saw great advances in both knowledge and treatment of animal diseases. Unfortunately these advances came at a cost both in terms of labour to apply the remedy and the cost of the medicine. A single vaccination became available for five diseases in sheep, as did newer and more effective drenches for control of intestinal worms. As stocking rates increase, other risk factors emerge: for example, stock water supply needs to be monitored for aspects of quantity and quality.

One major management change occurred during this period and this was to mate ewes in the autumn and lamb in the spring. When the Western District was first settled it was impossible to rear lambs unless they were born in the spring. Improved pasture based on clover and grasses made sufficient growth to support lactating ewes, and the lambs were then weaned onto spring growth. However, as the stocking rates were increased winter growth was often insufficient for the nutritional requirements of the ewes. Feeding out in the winter was often difficult or impossible due to water-logged paddocks. Spring lambing better provided for the nutritional requirements of the ewe but also required supplementary fodder, such as early cut hay or grain, to feed the weaners over the first summer.

Diversification is one method of reducing the dependency of a grazing property on sheep production. Most properties carry some cattle either as breeding or fattening enterprise. Lanark carried just under 1000 DSEs in cattle in 1964. An additional enterprise involved fattening Friesian steers. The proximity of Hamilton to the large concentration of dairy cattle to the south and southwest gave a ready supply of young male cattle, some of which are crossed with beef breeds. In 1968, following the drought some graziers boosted their stocking rate by purchasing Friesian heifer calves and having them contract-raised for 8 to 10 weeks before taking delivery and growing out the yearlings.

Some forms of diversification, while appearing to be lower yielding in financial terms, may have a role to play in property management. For example, where the labour resources are fully utilised an enterprise not related to livestock—such as farm forestry—may be an attractive alternative.

During the 1960s, increased awareness of environmental issues led to a greater understanding of the flow-on effects of many farming practices. This sustainability aspect, while not explicit, was evident in many of the recommended practices of the earlier period. It was widely accepted that good farming was good business. Property owners were aware that improving the pastures and developing an attractive landscape ensured future capital gains.

One aspect of Lanark was that the property was left in trust with the second generation of children to inherit. This was not uncommon in the Western District because at that time inheritance taxes (death duties) could impose a severe financial burden on the beneficiaries and by leaving assets in trust one generation could avoid their incidence. One disadvantage of this arrangement was that the person who held the property in trust could not borrow against the property assets. In addition, when the number of beneficiaries was large, the chance of anyone buying out all other beneficiaries became impossible unless there were large off-farm investments. Even in a case where only two beneficiaries were involved, funding the acquisition of 50 per cent of a property by one was financially difficult.

Farming is an exercise in capital formation. The manager of Lanark decided that expansion was the only method of creating a business of sufficient size to ensure the future viability of the Lanark property under one owner. While land was purchased, the returns from grazing livestock were insufficient to make rapid progress in repaying debt. The rate of return to capital in the grazing industry is traditionally low, and while the initial purchase of land appears manageable, appreciable extra capital is tied up in stock, which may represent income foregone if stock are bred rather than purchased. For this reason, farm businesses might choose to expand by leasing land and breeding up stock numbers rather than investing in both land and stock.

Lanark was managed to take advantage of business opportunities as they arose. For example a Border Leicester stud was carried on the property and provided another source of income. During this period a small flock of grey and black sheep were carried to cater for a growing market for their wool, destined for home spinning and knitting. A further enterprise was added during the drought when the price of young lambs was low. Lanark was able to buy a line of young sheep which, after drenching and feeding through the drought, were later sold for a substantial profit.

Mick Cook was a farm management consultant at Hamilton from May 1964 to December 1970. Following this period he joined the Commonwealth Public Service and carried out sheep surveys with the Bureau of Agricultural Economics. He joined the Department of Environment in 1974 and worked on a number of projects including environment legislation and a national study on soil conservation. Following a period employed as an economist in the Department of Housing in Canberra, he retired to Melbourne in 1993.

APPENDIX 3

Reflections on irrigation for sheep and wool production in the 1960s on the Basalt Plains

By Jim Murphy

Production under irrigation

I joined the Irrigation Branch of the Department of Agriculture in 1953, stationed at the State Research Farm, Werribee. The Research Farm had 500 acres (200 hectares) of summer-irrigated pasture of white clover, perennial ryegrass and cocksfoot. This pasture was utilised predominantly for dairy production, but some of the experimental areas were grazed by Corriedale and Border Leicester/Merino sheep. I was responsible for the conduct of experiments on the fertilizer requirements and grazing management of irrigated pasture.

Thirty sheep per acre could be carried on 40-year-old irrigated pasture in October, November and December: twenty per acre through January, February and March; and ten per acre from April to the end of September with no supplementary feeding required. Annual yields of 190 lb (180 kg) of wool and 400 lb (86 kg) of liveweight gain per acre were achieved. At 1965 prices the gross value of this production was about $130 per acre.

Expansion of sheep numbers

The 1950s saw sheep numbers almost double in the western region of the Basalt Plains. Hamilton vied with Yass in New South Wales for the title of 'Wool Capital of the World'. Expansion of the area of improved pasture replacing the less productive native pasture, greater use of superphosphate and increased fodder conservation provided opportunities to increase the stocking rate of pastures. At the Pastoral Research Station at Hamilton, for instance, sheep numbers increased from 1000 to 3000 between 1959 and 1963 and hay reserves from 3000 bales to 18 000 bales (*Weekly Times*, 30 April 2003, page 26).

Six DSEs per acre became a feasible stocking rate. The gross value of production at this stocking rate at 1965 prices was of the order of $60 per acre, compared with the irrigated pasture figure of $130 per acre. Eight DSEs per acre were achievable in good seasons but not sustainable without irrigation. Progressive graziers on the Basalt Plains considered the possible use of irrigation from on-farm storages of rainfall run-off to intensify and to provide greater security for their high stocking rate enterprises. The 1967 drought also provided a sobering experience for numerous high stockers.

Water harvesting

Low-cost systems of water supply and irrigation were essential for profitable irrigation for sheep and wool production. Fortunately, by the 1960s the principles of water harvesting had been established and publicised by Professor Hector Geddes of Sydney University. He sought out economical sites on the undulating country at Badgerys Creek near Sydney for the construction of gully dams to store run-off water. The dams needed to be sited high enough in the catchment to allow gravitational irrigation from them through a valved pipe through the embankment.

Economy of storage

The Storage/Excavation (SE) Ratio is a measure of the economy of a storage site. It is the ratio of volume of water stored to the volume of earth excavated for the bank. Geddes looked for sites with SE ratios of 5:1 to 6:1 for economic storage. At SE 6:1, water could be stored at a capital cost of $66 per acre foot.

The site requirements for a low-cost storage are:

1. a relatively flat gully gradient of 1 in 100 or flatter
2. a narrow neck in the gully for the embankment
3. sides of the storage basin widening out upstream of the embankment site.

The undulating topography of the Branxholme area was well suited for economic water storage and low-cost irrigation. Lake Cicely on Lanark, for instance, had an SE ratio of 12:1. The cost of water storage was $33 per acre foot in 1966. It was a remarkably low capital cost for its storage capacity of 60 acre feet.

Contour ditch irrigation

The Keyline System of land and soil sustainability was promulgated by P.A. Yeomans in the 1950s and developed on his two properties 65 kilometres west of Sydney. It had significant elements of water storage and low-cost irrigation.

Contour ditch irrigation, a slightly modified form of Keyline irrigation, provided a low cost method of irrigation for sloping country. Irrigation water is gravitated from a dam in an excavated channel or ditch on a slight gradient of about 1 in 1000. The channel is blocked at successive intervals causing water to overflow and irrigate the land below. The system can be imposed on an existing pasture, the only earthworks being the construction of the channel. This was often done with a rotary ditcher.

This system of irrigation is not new. It was introduced into Spain in the Middle Ages by the Moors. On several Spanish mission properties in San Antonio, Texas, crops were irrigated in the eighteenth century by this means. Water was diverted by gravity from a diversion dam on the San Antonio River into 'acequias', the Spanish term for channels. The system is still operational after three centuries. The Hanging Gardens of Babylon may well have been irrigated by similar means in the second century BC with water lifted from the Euphrates River by slave labour.

Irrigation management

Sheep and wool production is at the bottom of the list for returns from irrigation. Irrigated pasture utilised for city milk production, for example, at Badgerys Creek showed a gross return of $350 per acre with city milk at 50 cents per gallon. For irrigated pasture to be profitable for sheep and wool production its management had to be at a high level. The gross return was only $130 per acre.

The first requirement, once water storage and irrigation systems have been settled, is to introduce white clover and cocksfoot into the pasture to optimise its production in summer. Sod-seeding into an already improved pasture of sub clover and perennial ryegrass is practicable. If the area to be irrigated is still under native pasture it needs to be cultivated and properly sown down to an irrigation pasture mixture of white clover, cocksfoot and perennial ryegrass.

Irrigated pasture requires heavier dressings of fertilizers than dryland pasture. Some form of rotational grazing is desirable to prolong its productive life. Utilisation solely by grazing should be the aim. Cutting irrigated pasture for hay should be avoided. Regrowth after hay cutting is slow. Production of green feed may be reduced by as much as 35 per cent in the six weeks following hay cutting. Fodder conservation should be confined to the non-irrigated pastures on the property.

Mature sheep, two-tooth and older, fatten readily and their wool production is intensified when grazed for only 50 per cent of the time on irrigated pasture. Stocking rates of up to 25 per acre are possible through spring, summer and early autumn.

Irrigation provides the opportunity to have fat sheep available in autumn when store condition is the norm. It needs to be noted, however, that lambs do not fatten as readily on irrigated pasture as do mature sheep. Corriedale and Merino cross sheep are better suited to utilising irrigated pasture than are straight Merinos.

Irrigation also is not completely reliable. Dams did not fill in the 1967 and 1983 droughts. The dry years since 1998 have also restricted supplies of irrigation water.

A property needs to be close to full development before irrigation is introduced. Pasture improvement, fertilizer usage, grazing management and increased stocking rates yield better returns than irrigation. All of these aspects need to be intensified when irrigation is introduced. There is an increased labour component as well. Another aspect to be considered in the economic assessment of a proposed irrigation scheme is the loss of production from the area inundated by the dam.

Irrigation on Lanark

The Fentons toyed with the idea of irrigation even to the extent of contructing Lake Cicely, but did not proceed further. Was it an opportunity lost? Arguably yes, but there were sound enough reasons for their decision. It was a first-class irrigation site and 40 acres (16 hectares) could have been irrigated from the 60-acre-feet (74 000 cubic metres) storage.

However, John and Cicely were committed to raising a young family and to wresting a living, without hired help, from their 2000-acre (800-hectare) property. Irrigation would have required more intensive stocking, additional management commitments and a greater labour input. The cost of hired labour required would have greatly reduced the profit from irrigation. Restoration of Lanark's natural environment and diversification into farm forestry also loomed large on the Fentons' priorities.

Irrigation of woodlots was not a considered option for Lanark in the 1960s. In retrospect it would almost certainly have augmented the returns from farm forestry.

Lake Cicely has been a boon to Lanark in numerous ways even without irrigation. Indigenous planting around its perimeter and its combination of shallow and deep water are ideal for supporting a rich and varied wildlife. Stock water supply and fire protection are assured. The landscape quality of Lanark has been enhanced and its land value improved by the attractiveness of this body of water.

Present-day opportunities

The principles established for profitable irrigation for sheep and wool production are still basically valid. Prices of wool and meat are a far cry from what they were in the 1960s, but costs have also escalated.

The live sheep export trade to the Middle East has been a saviour for the sheep meat industry. It has given a boost to the returns from fattening sheep on irrigated pasture. The trade has expanded tremendously since its introduction in the 1960s. Originally confined to culled-for-age sheep, it has been a thriving market since the early 1970s for younger sheep in good condition, the type readily produced with irrigation. In the year 2000 live exports from Australia of 6 million sheep averaged $67 per head, a price undreamed of in the 1960s.

Storage of rainfall run-off on farms for irrigation is no longer the regulation-free procedure it was in the 1960s. A licence is now required for irrigation from a dam prior to its construction. A licence for construction of the dam may also be required. Restrictions may be placed on the size of the dam. In addition to licence fees there is also an annual charge of about $7 per megalitre.

Careful consideration of the pros and cons of intended irrigation schemes for sheep and wool production is still essential. The gross returns are greater than in the 1960s, but are eroded more by increased costs of all inputs especially fertilizer, labour and water charges.

APPENDIX 4

Lanark bird list (1956–96)

Compiled by Murray Gunn from his observations and records over this period and including additional sightings by John and Cicely Fenton.

The Lanark bird list follows the order published as the definitive taxonomic list of Australian birds by Christidis and Boles in 1994, and updated in 2008.

1956

Woodland species

Whistling Kite
Brown Falcon
Peregrine Falcon
Red-rumped Parrot
Blue-winged Parrot
Barn Owl
Yellow-rumped Thornbill
Red Wattlebird
Black-faced Cuckoo-shrike
Australian Magpie
Willie Wagtail
Australian Raven
Little Raven
Welcome Swallow

Wet grassland species

Black Swan
Australian Shelduck
Grey Teal
Pacific Black Duck
White-necked Heron
White-faced Heron
Australian White Ibis
Straw-necked Ibis
Brolga (disappeared until 1996)
Masked Lapwing
Latham's Snipe
White-fronted Chat
Magpie-lark
Fairy Martin

Grassland species

Stubble Quail
Black-shouldered Kite
Wedge-tailed Eagle
Nankeen Kestrel

Black Falcon
Banded Lapwing
Flame Robin
Brown Songlark
Richard's Pipit
Australasian Pipit

Aerial species

White-throated Needletail
Fork-tailed Swift

1956–76 arrivals

Woodland species

Diamond Dove
Brown Goshawk
Australian Hobby
Yellow-tailed Black Cockatoo
Galah
Long-billed Corella
Sulphur-crested Cockatoo
Musk Lorikeet
Purple-crowned Lorikeet
Crimson Rosella
Eastern Rosella
Budgerigar
Horsfield's Bronze-Cuckoo
Shining Bronze-Cuckoo
Pallid Cuckoo
Fan-tailed Cuckoo
Southern Boobook
Azure Kingfisher
Laughing Kookaburra
Superb Fairy-wren
Striated Thornbill
Brown Thornbill
Striated Pardalote
Eastern Spinebill
Yellow-faced Honeyeater
Singing Honeyeater
White-plumed Honeyeater
Little Wattlebird
Tawny-crowned Honeyeater
New Holland Honeyeater
White-naped Honeyeater
Golden Whistler
Grey Shrike-thrush
Olive-backed Oriole
Masked Woodswallow
White-browed Woodswallow
Dusky Woodswallow
Grey Currawong
Grey Fantail
Restless Flycatcher
Jacky Winter
Scarlet Robin
Rufous Songlark
Silvereye
Tree Martin
Red-browed Finch

Wetland species

Musk Duck
Australian Wood Duck
Pink-eared Duck
Australasian Shoveler

Chestnut Teal
Hardhead
Blue-billed Duck
Australasian Grebe
Hoary-headed Grebe
Great Crested Grebe
Darter
Little Pied Cormorant
Great Cormorant
Little Black Cormorant
Australian Pelican
Australasian Bittern
Great Egret
Little Egret
Eastern Great Egret
Nankeen Night Heron
Glossy Ibis
Royal Spoonbill
Yellow-billed Spoonbill
Swamp Harrier
Purple Swamphen
Lewin's Rail
Buff-banded Rail
Baillon's Crake
Australian Spotted Crake
Black-tailed Native-hen
Dusky Moorhen
Eurasian Coot
Black-winged Stilt
Red-capped Plover
Double-banded Plover
Black-fronted Dotterel
Red-kneed Dotterel
Common Sandpiper
Common Greenshank
Red-necked Stint
Sharp-tailed Sandpiper
Curlew Sandpiper
Whiskered Tern
Silver Gull
Clamorous Reed-warbler
Australian Reed-warbler
Little Grassbird

Wet grassland species

Cape Barren Goose
Golden-headed Cisticola

Grassland species

Spotted Harrier

1976–96 arrivals

Woodland species

Common Bronzewing
Little Eagle
Gang-gang Cockatoo
Rainbow Bee-eater
White-browed Scrubwren
Spotted Pardalote
White-eared Honeyeater
Varied Sittella

White-winged Triller
Rufous Whistler
Rufous Fantail
White-winged Chough
Pink Robin
Eastern Yellow Robin
Bassian Thrush

Wetland species

Magpie Goose
Plumed Whistling-Duck
Freckled Duck
Intermediate Egret
Spotless Crake
Banded Stilt
Wood Sandpiper

Wet grassland species

Brolga (returned 1996)

Post-1996 arrivals (by Murray Gunn)

Since 1996 my observations of the birds at Lanark have been much less frequent, and as John and Cicely also left the property during this period there has been less opportunity to record any new species that may have appeared during this time. John and Cicely's son David lets me know of any new sightings.

It could be argued that most of the species likely to be found at Lanark had been recorded by 1996. However, there are some species that have not been recorded that would have occurred at some stage. These include Tawny Frogmouth, Pied Cormorant, Cattle Egret, Collared Sparrowhawk, Red-necked Avocet, Painted Button-quail, Sacred Kingfisher, White-throated Treecreeper, Yellow Thornbill, Brown-headed Honeyeater, White-bellied Cuckoo-shrike, Satin Flycatcher and Mistletoebird.

In spite of the relatively little time spent on bird observation since 1996, it does appear that bird species and numbers have declined since the 1976–96 period.

Some reasons for the decline are obvious. The ten-year drought with its effect on the wetland has had a marked effect on water birds occurrences. If and when the drought breaks, as it did partially in 2009, the wetland birds will return.

The decline in woodland species is more complex. Many woodland species, particularly in southeastern Australia, appear to be in decline, and that is a major concern for those who study avifauna. Once again the drought has been responsible for conditions not conducive to optimum breeding potential, so fewer young are produced to keep the population stable.

It is also possible that the effect of European settlement, with its clearing and fragmentation of the landscape, has had a delayed effect on many bird species. This is in spite of the work put in by the Fentons and, in more recent years, many other landowners in re-creating woodland habitat and wildlife corridors.

One can only hope that with these efforts to improve the environment and the cessation of further woodland and forest clearing, the decline of many species will be halted and hopefully, even reversed.

The five graphs on the following pages chart the change in numbers of bird species (and in the relative size of their habitats) at Lanark from 1840, to 1956, when John Fenton took over the property, and in ten-year increments thereafter. These graphs were created by Elizabeth Jacka, who worked closely with Murray Gunn, with input from Graham Pizzey to estimate the bird population at Lanark at the time of white settlement.

1840

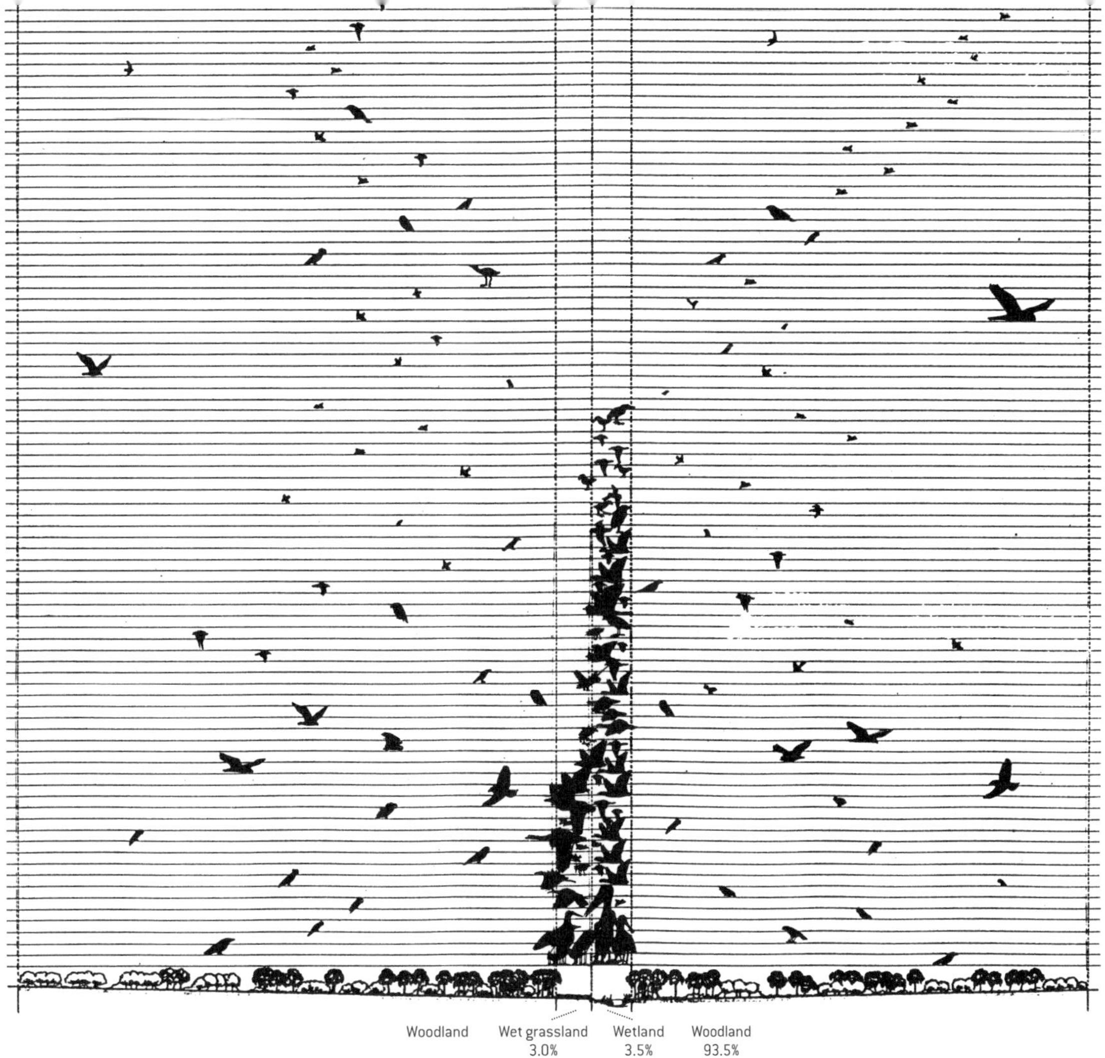

1956

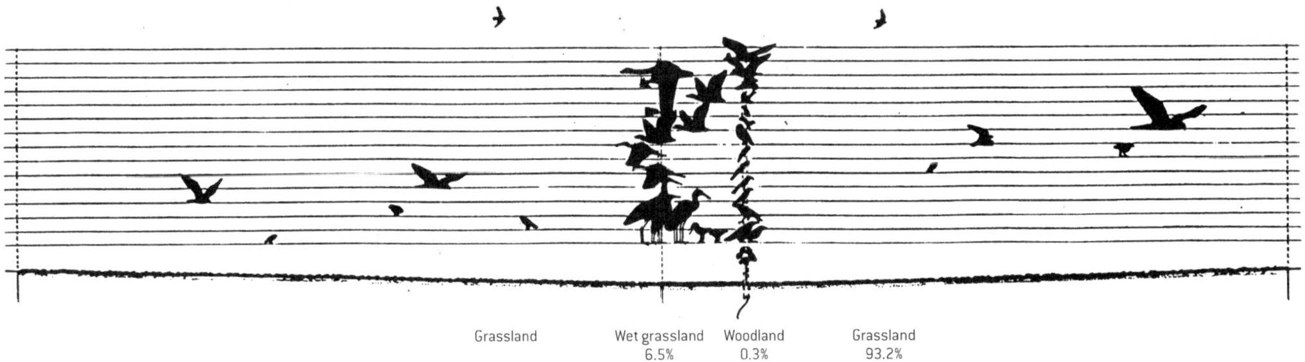

1966

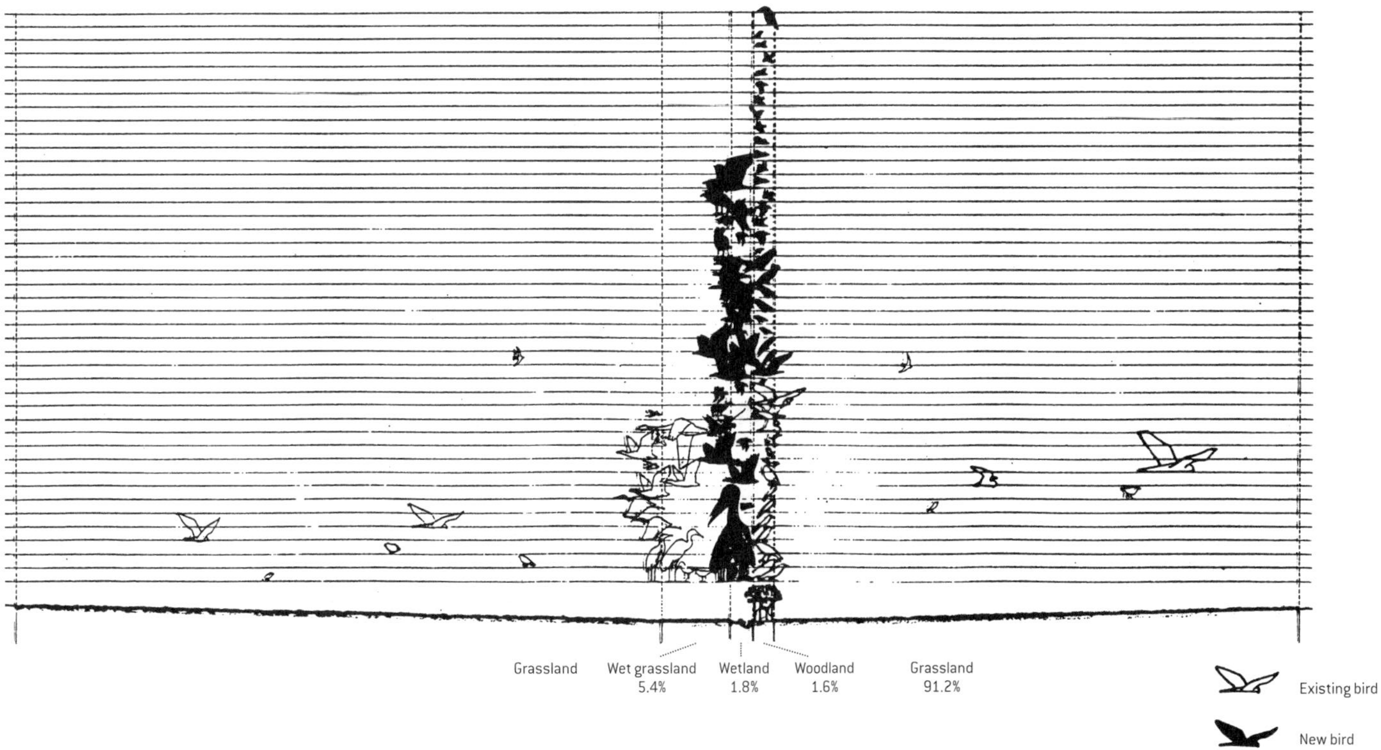

Grassland
Wet grassland
5.4%
Wetland
1.8%
Woodland
1.6%
Grassland
91.2%
Existing bird
New bird

1976

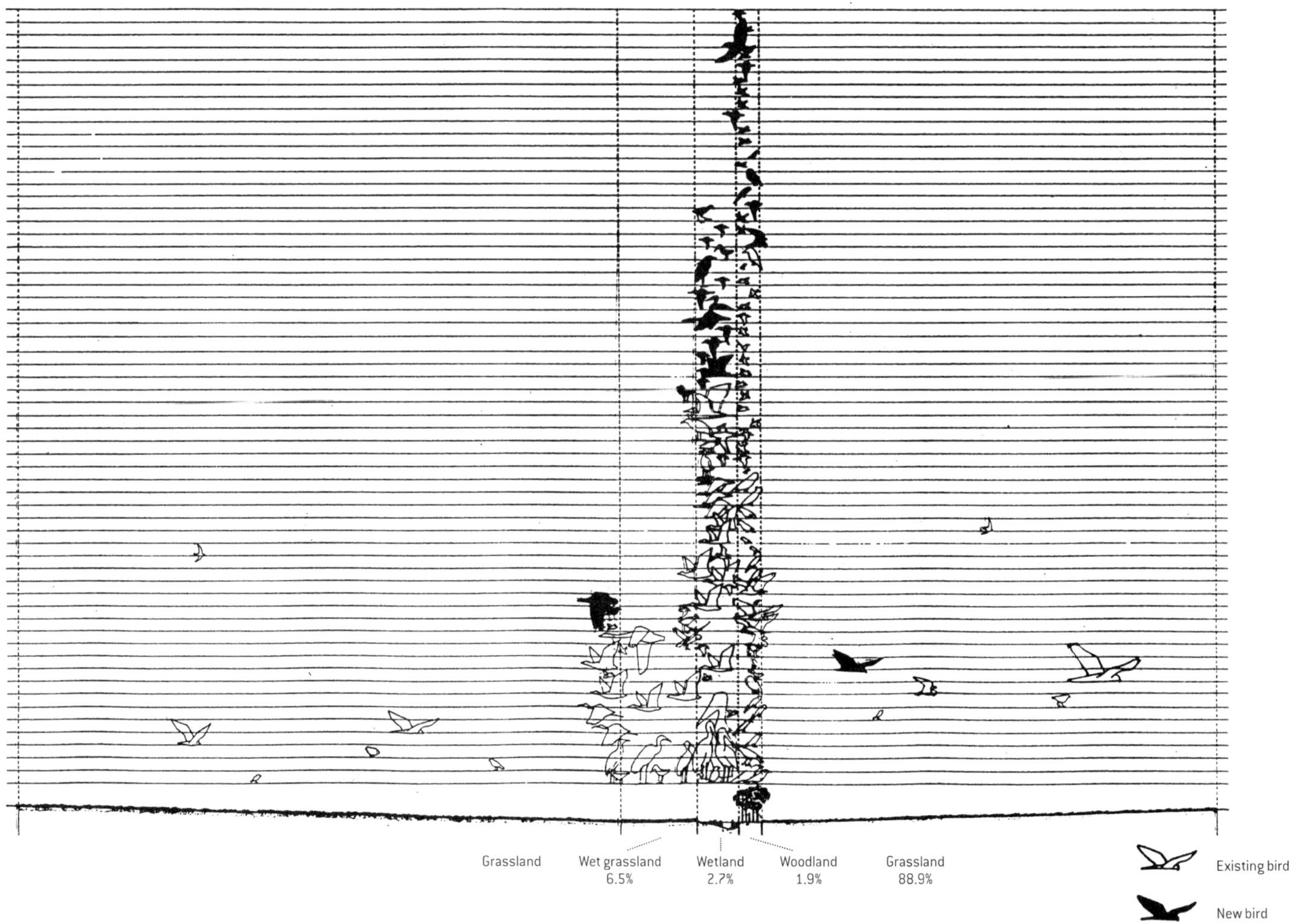

Grassland
Wet grassland 6.5%
Wetland 2.7%
Woodland 1.9%
Grassland 88.9%
Existing bird
New bird

1986

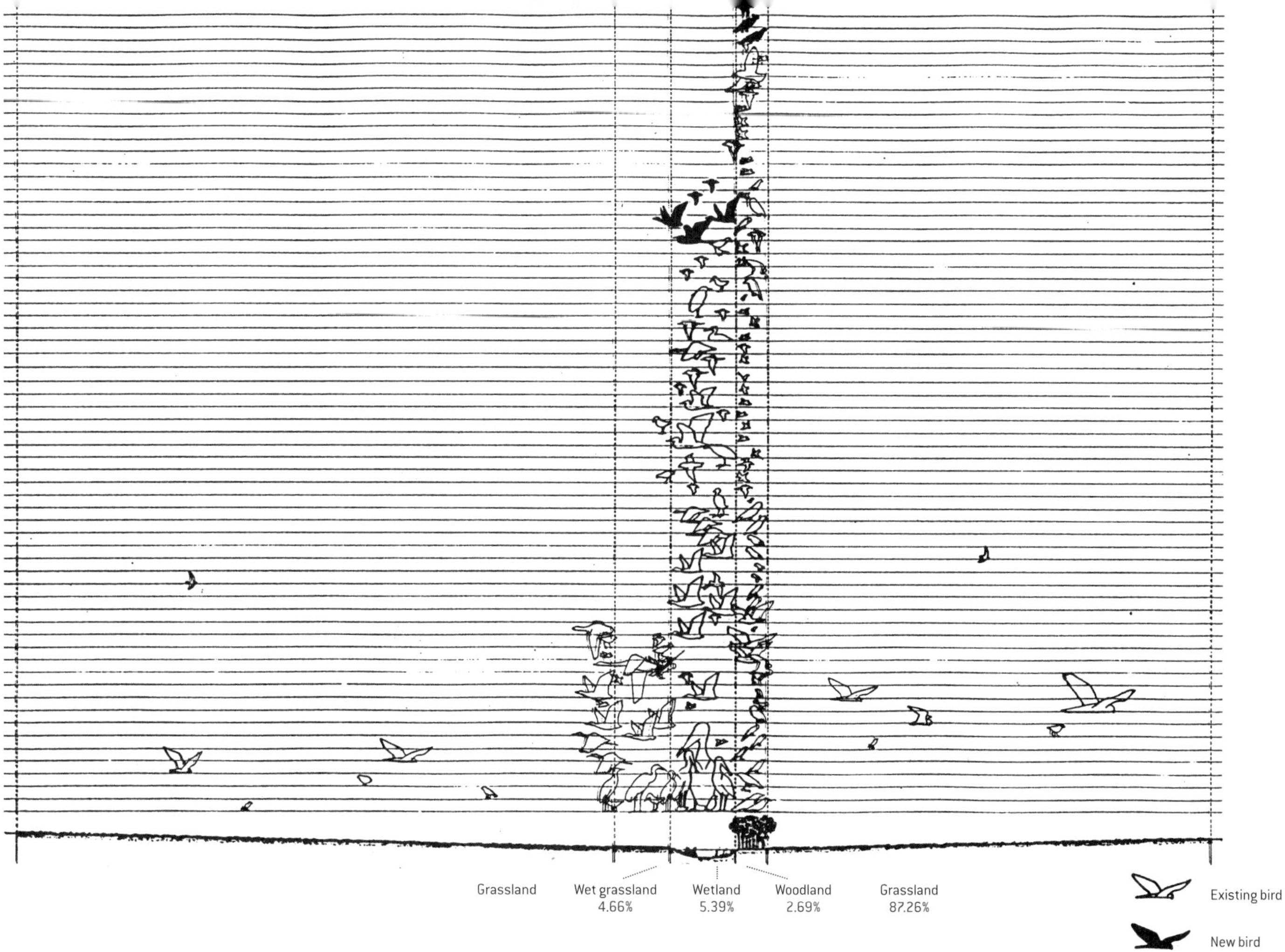

Grassland
Wet grassland 4.66%
Wetland 5.39%
Woodland 2.69%
Grassland 87.26%
Existing bird
New bird